The Animals of New Zealand

An Account of the Dominion's Air-Breathing Vertebrates

FREDERICK WOLLASTON HUTTON
JAMES DRUMMOND

CAMBRIDGE
UNIVERSITY PRESS

CAMBRIDGE UNIVERSITY PRESS

Cambridge, New York, Melbourne, Madrid, Cape Town,
Singapore, São Paolo, Delhi, Tokyo, Mexico City

Published in the United States of America by Cambridge University Press, New York

www.cambridge.org
Information on this title: www.cambridge.org/9781108040020

© in this compilation Cambridge University Press 2011

This edition first published 1923
This digitally printed version 2011

ISBN 978-1-108-04002-0 Paperback

CAMBRIDGE LIBRARY COLLECTION

Books of enduring scholarly value

Life Sciences

Until the nineteenth century, the various subjects now known as the life sciences were regarded either as arcane studies which had little impact on ordinary daily life, or as a genteel hobby for the leisured classes. The increasing academic rigour and systematisation brought to the study of botany, zoology and other disciplines, and their adoption in university curricula, are reflected in the books reissued in this series.

The Animals of New Zealand

Frederick Wollaston Hutton (1836–1905) was a geologist and a supporter of Darwinian theory. He emigrated to New Zealand in 1866, became Professor of Biology at Canterbury College, and won awards both in Britain and Australasia for his work on the natural history of New Zealand. He published scientific papers on biology and zoology as well as geology and, with James Drummond, wrote two popular works, *Nature in New Zealand* (1902) and *The Animals of New Zealand* (1904). The latter was extremely successful. It was revised and expanded the following year, and this fourth edition was published in 1923. The book focuses on native vertebrates, so the bulk of the content relates to birds, of which Hutton had published a catalogue in 1871. It also describes marine mammals, reptiles, and bats, and gives brief coverage to introduced species. There are 154 illustrations, and indexes of Maori, English and scientific names.

Cambridge University Press has long been a pioneer in the reissuing of out-of-print titles from its own backlist, producing digital reprints of books that are still sought after by scholars and students but could not be reprinted economically using traditional technology. The Cambridge Library Collection extends this activity to a wider range of books which are still of importance to researchers and professionals, either for the source material they contain, or as landmarks in the history of their academic discipline.

Drawing from the world-renowned collections in the Cambridge University Library, and guided by the advice of experts in each subject area, Cambridge University Press is using state-of-the-art scanning machines in its own Printing House to capture the content of each book selected for inclusion. The files are processed to give a consistently clear, crisp image, and the books finished to the high quality standard for which the Press is recognised around the world. The latest print-on-demand technology ensures that the books will remain available indefinitely, and that orders for single or multiple copies can quickly be supplied.

The Cambridge Library Collection will bring back to life books of enduring scholarly value (including out-of-copyright works originally issued by other publishers) across a wide range of disciplines in the humanities and social sciences and in science and technology.

FREDERICK WOLLASTON HUTTON

Born 16th November, 1836.   Died 27th October, 1905.

THE
Animals of New Zealand

AN ACCOUNT OF THE DOMINION'S AIR-BREATHING VERTEBRATES

BY

Captain F. W. HUTTON, F.R.S.

AND

JAMES DRUMMOND, F.L.S., F.Z.S.

(FOURTH EDITION: REVISED AND ENLARGED)

White-Fronted Tern

Auckland, Christchurch, Dunedin, and Wellington, N.Z.
Melbourne and London
WHITCOMBE AND TOMBS LIMITED
1923

CONTENTS

CONTENTS 7

8 CONTENTS

LIST OF ILLUSTRATIONS

PREFACE TO THE FIRST EDITION

In this work we have endeavoured to combine popular information with the purely scientific, and have intermingled stories of quaint habits and characteristics with accurate descriptions of all the animals dealt with.

Our object has been to publish a volume that will be useful to naturalists, and at the same time interesting to the general public. The first consideration has been accuracy; the second, sufficient information in small space. To attain this end, we have drawn largely on the works of others. Chief among these is the late Mr. T. H. Potts. He knew our fauna before its destruction had been fairly begun, and he devoted stupendous energy to the study of the habits of our native birds. As a result of his labour of love, New Zealand possesses unique records in its literature on natural history. The main part of his observations is embodied in stray articles in the Christchurch newspapers, in the *Transactions of the New Zealand Institute,* and in a little book entitled *Out in the Open.* No work dealing adequately with the natural history of this colony would be complete without quotations from Mr. Potts's writings; and as Mrs. Potts has kindly placed all his publications at our disposal, we have introduced into this volume some of the best of them. Where large and important collations have been made, we have indicated the source or the writer; but we have found it impracticable to do so in regard to all quotations, as we have had to take a few lines here and a few there, and to weld them together in order to make a connected whole. We have exercised much care in taking extracts. A portion of the popular part of the information was published as a series of copyright articles in *The Lyttelton Times* in 1903; but these have been revised and added to.

On the scientific side, the present volume may be considered as the third edition of the *Descriptive Catalogue of the Birds of New Zealand*,* much enlarged, with an account of the dominion's mammals and reptiles, as well as its birds. It was first published in 1871, and, in 1888, a second edition, with illustrations, was edited by Sir Walter Buller. Both editions have been out of print for many years.

A large portion of the volume represents original research, and a great deal of the information is now published for the first time.

We have supplied as many illustrations as possible. Most of them are reproductions of photographs of coloured plates, and it is necessary to state that blue colours appear white, and yellow dark. This is noticeable in the wattles of the crows and the skin on the faces of the cormorants, in the blue feathers on the wings of the parrakeets on the head of the bell-bird, and in the yellow on the bills of the mollymawks and the crests of the penguins. In connection with the illustrations, we have to thank Mrs. Potts for the excellent portrait of Mr. Potts. Also the Director of the British Museum, the Council of the Zoological Society of London, the British Ornithologists' Union, the proprietors of *Nature*, Sir Walter Buller, Dr. R. B. Sharpe, Dr. P. L. Sclater, the Hon. W. Rothschild,† Professor J. H. Scott, and Mr. F. E. Beddard, for permission to reproduce illustrations published by them, and also Dr. L. Cockayne for the photograph of the nesting albatross on page 271. All the nests are from specimens in the Christchurch Museum. These, as well as almost all the other illustrations, have been reproduced from photographs taken by Mr. W. Sparkes, Taxidermist to the Museum. As far as possible, we have gone to classical works for illustrations, thinking that this will add to the value of the book.

In dealing with the birds, we have omitted the wanderers; but every bird that belongs to New Zealand is described. The list of the dominion's avifauna, probably, is now closed, the

---

*By F. W. Hutton, published by the Government Printer, Wellington, New Zealand, 1871.
†Now Lord Rothschild (1921).

last named being that of the Bounty Islands shag, which was added in 1901, and makes a total of about 190 species.

In the measurements, the length of the wing means the length from the flexure at the wrist-joint to the tip of the longest feather. The tarsus is what is ordinarily called the leg; it lies between the drumstick and the foot.

PREFACE TO THE FOURTH EDITION

The first edition of this work was published in 1904. It was much more successful than Captain Hutton or I, or even the publishers, had anticipated, and a second edition was found to be necessary in the following year, 1905. Captain Hutton was then suffering from a serious illness. He went to England for a change, and was not able to take any part in the work of seeing the second edition through the press. Before he left New Zealand, however, we discussed several alterations to the first edition and decided to make some additions. A copy of the second edition reached him in England on September 19th, 1905, and he wrote to me on that date expressing his appreciation of the work. About five weeks later, October 27th, he died at sea on his way back to New Zealand.

In this edition I have made many alterations and additions. Archdeacon Williams, of Gisborne, has kindly revised the list of Maori names of birds, and, on his advice, I have changed a large number, giving some names that are more generally accepted than those previously assigned to birds in this book. Unfortunately, I could not find space for Archdeacon Williams's complete list of Maori names. He has published an excellent paper on the subject in *The Journal of the Polynesian Society* for December, 1906, No. 4, volume xv. It supplies all the names the Maoris are known to have applied to native birds. In one case, the bell-bird (*Anthornis melanura*), he gives no fewer than twenty-six names.

For many years ornithologists were under the impression that New Zealand's birds were rushing headlong to destruction, and that they would soon pass completely away. The utter extermination, in a flash of time, as I may say, of one of God's brightest and most harmless beings, which has been represented in this life for ages, is a regrettable incident in the world's history, and it is not

16

surprising that naturalists and scientists should commiserate with New Zealand in the loss that seemed to be impending. I am glad to be able to sound a brighter note. Some time ago, by the courtesy of the Agricultural Department, I had thousands of circulars sent to all parts of the dominion. These circulars contained questions in regard to our birds' present position. When they were returned to me, I found that almost all the birds had been accounted for except one. The missing bird is the native quail (*Coturnix novae-zealandiae*), which fell in thousands before the great grass fires that swept through the land as settlement advanced.

There is no convincing evidence that any New Zealand bird, except the Stephen's Island wren and the native quail, has been exterminated by the European inhabitants of this country. Although we regret the needless destruction that has taken place, we may find some satisfaction in the knowledge that the position is not nearly as bad as we thought it was. Our birds may still be seen wherever the ancient forests stand. As long as large tracts of forests are left as sanctuaries and scenic reserves, we need not have much fear that our avifauna will be completely lost. Mr. H. G. Ell, for many years a member of the House of Representatives, has done New Zealand a great service in urging that the forests should be preserved and the birds protected, and all that he asked was willingly done by Sir Joseph Ward, the Hon. W. F. Massey, and other Ministers. Naturalists in all parts of the world will be grateful for a Scenery Preservation Commission, which went through the country and reported upon sites that ought to be preserved, and also to Mr. S. Percy Smith, chairman of the Commission, and Mr. W. W. Smith, the secretary, and other members for the enthusiasm with which they undertook their congenial duties.

In the second edition, "Pelorus Jack," the famous cetacean that followed steamers through Pelorus Sound, was classified as a goose-beak whale (*Ziphius cavirostris*). As he was protected by the Legislature as a Risso's dolphin (*Grampus griseus*), I took him out of the former species. The first Order-in-Council

B

under which he was protected was signed by His Excellency the Governor, Lord Plunket, on September 26th, 1904, and was published in the official "Gazette." The order was issued under the Sea-fisheries Act of 1894, which empowers the Governor in Council to make regulations protecting any fish. The last report of "Pelorus Jack" was in November, 1916. He has not been seen from that date up to the present time, April, 1922.

It is with deep regret that I have prepared later editions without Captain Hutton's assistance. We spent many hours of hard work studying natural history, and I feel that if he had lived longer we would have done more together. Many years of his life were devoted to scientific research, for which he was splendidly equipped by his powers of observation and judgment and his great industry.

He was the second son of the Rev. H. F. Hutton, and was born at Gate Burton, Lincolnshire, on November 16th, 1836. He was educated at Southwell and at the Naval Academy at Gosport. As he was over age at the time of his nomination, he could not enter the Royal Navy, but he served for three years in the India mercantile marine. He then entered King's College, London. He became an ensign in the 23rd Royal Welsh Fusiliers in 1855, and was made lieutenant in 1857 and captain in 1862. After serving in the Crimea in 1855 and 1856, he went to India, and was present at the capture and relief of Lucknow and at other engagements. In 1860, he returned to England, and six years later he came to New Zealand with his family. He was appointed Assistant Geologist to the New Zealand Geological Survey in 1871, Provincial Geologist of Otago and curator of the Otago Museum in 1873, and Professor of Natural Science at the Otago University in 1877. In 1880 he was appointed Professor of Biology and Geology at Canterbury College, Christchurch, and in 1893 he became curator of Canterbury Museum, a position which he held until his death.

His services to his country have been recognised by the erection of a tablet in Canterbury College, and by the "Hutton Memorial Medal," which was struck by the Governor of the New Zealand Institute, with the assistance of the New Zealand Government. His most lasting memorial, however, will be found in his literary work. He edited the *Index Faunae Novae Zealandiae* for the Philosophical Institute of Canterbury, wrote *A Class-book of Elementary Geology, Darwinism and Lamarckism Old and New,* and *The Lesson of Evolution,* contributed largely to scientific journals in different parts of the British Empire, prepared thirteen descriptive catalogues and geological reports for the New Zealand Government, and published more than a hundred papers in the *Transactions of the New Zealand Institute* and other periodicals in Australasia. He had studied many branches of science in New Zealand, and investigators in the future will be grateful to him for the work he did.

J. D.

Christchurch, New Zealand.
April, 1922.

THE
Ånimals of New Zealand

INTRODUCTION

New Zealand contains a remnant of the population of a
continent that existed long before mammals overspread the
earth. That population was added to as the ages passed, fresh
colonists dropping in from time to time, mostly from Australia.
At last, not many hundreds of years ago, there came Man.
The coming of the Maori was as the shadow of death to a large
section of the original inhabitants; but it was only a prelude
to the great transformation wrought, as the Europeans, who
followed, swept the land with the besom of destruction. The
dramatic events of that time are dealt with a few pages further
on; we wish to point out here some of the effects of the
dominion's long isolation, and to deal with a few of the
peculiarities that have made New Zealand famous.

Our animal life is full of contrasts and surprises. The
manifold eccentricities of our fauna are so striking that some
naturalists would like to see New Zealand constituted one of
the great Zoological Regions of the globe. Although Dr. A. R.
Wallace refused to grant it this honour in his *Geographical
Distribution of Animals,* and made it a sub-region of Australia,
everybody admits that it has many claims to distinction, on
account of those things it lacks, as much as on account of those
it possesses.

Our birds are noted for flightlessness and songlessness and
dullness of plumage. Yet some of them remain on the wing
during extraordinary flights over vast stretches of water and

21

some are classed amongst the best songsters on earth. Sir Joseph Banks has described with enthusiasm the day-break chorus of the bell-birds and the tuis and their smaller companions, together with the flute-like notes of the saddle-back. Many of the early settlers also were delighted with this Song of Dawn, which, however, can be heard now only in places far removed from settlement.

Some of our birds, besides being peculiar to New Zealand, possess characteristics that single them out for attention. Among these is the wry-billed plover, which is the only known bird with a bill bent to one side.

We are entirely devoid of four-footed beasts that divide the hoof or chew the cud. Our only land mammals, with the exception of the seals, are two bats, and one of them is found in no other country. Our "creeping things of the earth" consist almost entirely of lizards, there being absolutely no snakes here. Even lizards are represented by only fifteen species. But among the reptiles is the famous tuatara. It is not a true lizard, as, in structure, it shows an affinity to the crocodile, and its ribs have bird-like characteristics. It stands apart from all the rest of its class, and if ancient lineage, combined with unchanged habits, mark the aristocrat, it is the most aristocratic animal in the world. We are even worse off for amphibians than for land mammals and reptiles.

THE BEGINNING.

Towards the close of the age in the world's history called the Cretaceous Period, New Zealand was a small group of islands, with a very scanty flora and fauna; but later on, very early in the Tertiary Era, it was gradually elevated until it attained almost continental dimensions, stretching away north through New Caledonia and Fiji, and joining the mainland at New Guinea. The land was covered with such vegetation as ferns and forest-trees. So far as animal life was concerned, however, it was the abomination of desolation, as the forests contained no birds, and the fern-lands no lizards except the ancient tuatara.

Birds had only lately come into existence in the Northern Hemisphere, but now that New Zealand was joined to the mainland, they moved south and colonised it. There were kiwis, crows, thrushes, starlings, robins, wrens, parrots, rails, ducks, penguins, and other birds, as well as the lizards and the frog. Towards the close of the Eocene Period, the northern land sank. New Zealand was then isolated, and it has remained so ever since.

Communication with the Outside World.

Yet it was visited every year by migratory birds from the north. Cuckoos, plovers, and sandpipers crossed the waters, and returned again to the place from which they had come. They brought no news of what was passing in the outside world, or, at any rate, they imparted none to the inhabitants of this country. More colonists dropped in occasionally, and took up their permanent residence here. There were the quail and the hawk, and, later on, the harrier, the grey duck and the shoveller, the white heron and the blue heron, and the black-footed shags, all from the north, while the gulls, the terns, the albatrosses, and the pink-footed shags came from the south.

The migratory birds still maintain their visits, coming regularly at the appointed time, and departing again to other countries. It is hard to say what attractions we offer to them, and harder still to define the faculty that directs them on their long and weary flights, sometimes from Siberia, as in the case of the godwit and the knot.

Bird migrations also take place in the Northern Hemisphere, where, obviously, migrants pass from one country to another on account of food supplies, which, for insect-eating birds, differ much in the summer and the winter. But this is not the cause of all the migrations. Swifts migrate in Central America, while swallows remain there all through the year. Ducks do

not leave their winter quarters for want of food; and our
godwit would find just as much food on our shores in the
winter after it has left, as in summer, when it first comes.
Our bronze and long-tailed cuckoos migrate from New Guinea.
In that country three or four species of bronze cuckoos are
resident. They evidently find food there all the year round,
and what they could do in this respect could also be done by
the migrating species, which visit New Zealand and Tasmania.
With us, the cuckoos are harbingers of summer, and, even in
summer, insects are much less plentiful here than in New
Guinea or Australia. It is clear therefore, that the cuckoos are
not attracted to us by an abundance of insect food.

Their migrations may probably be explained by the theory
that a habit has been formed by resorting each year to the
same breeding place. The birds do not like to break away from
the old-time custom, and it may be that they are attacked by
home-sickness. Perhaps the cuckoos, after living for some
months in a distant land, cherish the sentiment expressed by
the coloured man, whose heart longed for the old folks at home,
and the cabin on the Mississippi shore.

In respect to the *Limicolae,* or shore birds, such as the
plover, the godwit, and the sandpiper, it is likely that they
return annually to the old feeding ground of their forefathers,
and that this habit also has become an instinct.

In the Northern Hemisphere, migratory birds, as a rule,
follow the land. Some of them, however, have to cross the
Mediterranean Sea, others the North Sea, and others the
English Channel; and many shore birds pass from island to
island in the Malay Archipelago. But the boldest flight on
record is to New Zealand and the Chatham Islands, probably
from New Caledonia, a distance of a thousand miles or more.
The question may be asked, "How do birds know that they
will find land at the end of their long voyage?" They do not
fly at random; they go voluntarily; they must know there is
land ahead. The information is not given by "stragglers,"
which have been lost from a migrating flock, and have gone
to some other country, as these do not return to start new lines

of migration, but either establish themselves as residents in the strange country, or perish. Nor do wanderers leave behind them traditions of new routes. The only possible explanation seems to be that the migrants are following old land lines, which were quite clear at one time, but have now disappeared. The shore-birds follow the old shore-lines, and the land-birds the old land-line. Migration must have commenced when the two lands, which are now separated, were contiguous, or nearly so. In no part of the course would an island be so far off at first as to be invisible from an adjacent one. Then the land must have gradually sunk. But the force of habit, handed down from generation to generation, probably maintained the migration, until it became an instinct. During the life of each bird, the changes would be too small to be perceptible, and only after many years had passed would the migrants find that they were flying over a trackless ocean.

The faculty of returning home, whatever the nature of it may be, is not uncommon in the Animal Kingdom. Bees and ants and many other insects possess it to a small degree. Sea-snakes and turtles return to the same place to breed, although, during their absence from land, they must have swum many miles in many different directions. The faculty is also possessed by penguins, petrels, and other birds. Several seals undertake long oceanic migrations. As is well known, some domesticated animals return to their homes after having been taken long distances away; and, finally, savages, after having followed their quarry for several days, find their way home again through dense forests. The faculty, however, is not unerring, even with migratory birds. In the course of migrations, large numbers of "stragglers" lose their way; many perish at sea, and perhaps none regain the route after having lost it.

Before passing on to other questions, it may be asked, "Why should some of the shore birds and the two cuckoos migrate to New Zealand, while the swallows, which are certainly quite as capable of undertaking the journey, do not come?" Possibly the answer may be found in the geological history of birds.

The evidence is of a negative palæontological character, and it must be used with great caution; but it seems probable that the godwit and the cuckoo migrated to New Zealand at a time when there were no swallows in existence, and that the original land bridge had been completely broken down before the first of the swallows arrived in Australia from Asia. We may therefore suppose that migration to and from New Zealand commenced in the Eocene Period, when the land stretched away north-west to New Guinea, a time when all New Zealand was joined to the mainland.

A HAPPY FAMILY.

All along the line the effects of isolation are very marked. The absence of land mammals has been already touched upon. The presence of a great many species of birds peculiar to the country is another feature. As time passed, the birds that had come down to these parts found they possessed a land of surpassing goodness. It was free from drought and other disasters with which the faunas of many countries are beset; and abundance of food was easily obtained. Moreover, there were practically no natural enemies. The birds, as a whole, were a happy family. They had their petty quarrels and bickerings, but there was no common foe greatly to reduce their numbers. Life was too easy for them; so many first neglected, and then lost, the power of flight, and dropped into an indolent way of doing things, which became their undoing.

Excluding the birds from the Chatham and the Auckland Islands, we possess, of the first six orders of land-birds, only forty-five species, and no fewer than about thirty-eight of these are endemic. These forty-five species have been referred to thirty-one genera, nineteen of which are found nowhere else, and these thirty-one genera belong to twenty families, two of which, *Stringopidae*, represented by the kakapo, and *Xenicidae*, represented by the wrens, are peculiar to New Zealand. Our two owls are also peculiar, and one of them, the laughing owl, belongs to a genus found in no other country; these facts are specially notable, seeing that owls are very widely spread.

Our parrakeets, although belonging to a genus also found in Polynesia, differ much from those of Australia. It is remarkable that we have no representatives of the cockatoos or the grass-parrakeets, which are common in Australia and Tasmania, although our climate is quite suitable for them. This shows that the Eocene continent was very poorly off for land-birds.

Waders are more widely spread than the birds of any other order, and some of them are almost cosmopolitan; but even here the isolated character of our fauna is noticeable; for of thirty-two species, belonging to twenty-four genera, ten species and four genera are found nowhere else. The most marked feature in respect to the order is the presence of the wood hen, a genus of rails quite unable to fly. Species of closely related genera are also found in Lord Howe Island and New Caledonia. *Notornis hochstetteri*, which resembles both our own swamp hen and the *Tribonyx* of Tasmania and Australia, is another notable endemic rail. New Zealand is the only country in the world inhabited by two species of stilt-plover, neither of which is found elsewhere. Among the water-birds, cormorants are largely developed, as we possess fifteen species, twelve of which are endemic. No other country in the world possesses so many of these birds. We have two species of gulls found nowhere else, and this is a peculiarity of which few countries can boast. The most remarkable circumstance connected with our ducks is the presence of a species of *Fuligula*, a genus found in neither Australia nor Africa, but belonging properly to the northern parts of America, Europe and Asia, although one species occurs in South America.

New Zealand, together with the neighbouring islands, may be looked upon as the headquarters of the penguins, as all the genera except one are found here. Besides this, the oldest penguin known is from the rocks of New Zealand; and this country is probably the centre from which these birds dispersed.

Taking our fauna as a whole we find that the elements represented are Australian, Melanesian, European, Antarctic, and South American, the last being the weakest. But our birds show only three elements, namely, Antarctic, Melanesian, and Australian.

During isolation, some of the animals, notably the short-tailed bat, the huia, the thrush, the kakapo, the rails, and the extinct moas, altered a great deal, while others, such as the fern bird, the warbler, the tits, and the swamp hen, altered very slowly; but there was a tendency in all towards losing their powers of flight. Albinism is noticeable among the land-birds, but it does not remain constant. Melanism, or dark-coloured plumage, is very pronounced. Many of the birds during the long isolation have gradually changed until some of them have taken on almost a jet black plumage. There are the huia, the tui, two robins, a fantail, the red-bill, the black stilt, and a black penguin. This is a long list of black or dark-coloured birds, considering the smallness of the total number in the fauna. We merely point out this fact in passing, and offer no theory as to the cause.

THE CHANGE.

So the fauna, retaining its peculiarities, and developing its wonderful specialisations, grew up side by side with the flora, which is hardly less remarkable. They walked hand-in-hand, and were indivisible. The trees and shrubs yielded their fruits, and the birds rested in peaceful valleys and filled the land from end to end.

About 500 years ago, the Maoris came as the heralds of the change. They may have exterminated the moas, hunting and killing the great birds and eating the eggs. The white heron nearly succumbed in the North Island. A dog and a rat were introduced by the Maoris, but the former was excessively lazy and the latter timid and harmless, so neither had much effect on the fauna. The dog has disappeared entirely, but the rat is still with us. Though different kinds of birds were caught and snared by the Maoris, very little damage, on the whole, was done by them.

It was with the advent of Europeans that destruction began in earnest. It seemed as if they had been commanded to destroy the ancient inhabitants. Men went forth with slaughtering weapons in their hands, and the overflowing scourge of devouring fire was sent throughout the country, till a great part of the flora was consumed, and the birds had to seek food in remote places.

The first European animals were introduced by Captain Cook. On his second voyage, in 1773, he let loose three pigs in Queen Charlotte Sound. His motive was purely philanthropic. He made the Maori to whom he gave the pigs promise not to kill them. "If he keeps his word, and proper care is taken," the navigator wrote in his diary, "there are enough to stock the whole island in due time." The Maori did keep his word, and it was not long before the country was completely stocked with "Captain Cooks," as they were called by the settlers. They afforded food and sport for the Maoris, and also for the adventurous Europeans who lived in the colony in its early civilised days. They also took part in the war of extermination, and it was probably owing to their depredations that the tuatara was banished from the mainland. Less than a hundred years after the first liberation, the settlers in many parts looked upon the pigs as a scourge, as they killed the lambs. Cook's prediction was fulfilled too well. The pigs, becoming quite wild, retreated from the sites of civilisation, but in uninhabited valleys they congregated in vast numbers. Dr. Hochstetter, writing in 1862, after a visit to the colony, states that experienced pig-hunters sometimes took contracts for the suppression of the pigs, and that in twenty months three men, on an area of 250,000 acres, killed no fewer than 25,000, and pledged themselves to kill 15,000 more.

An inkling of the full effects that were to follow the advent of civilisation was given when the whalers visited these waters and established stations on the shores. Their special mission was the destruction of the whales, but they made their presence felt in other ways, notably by the introduction of dogs, cats, and rats, before which many of the flightless birds fell easy victims.

Then the settlers arrived with fresh supplies of domestic animals and deadly fire-arms. By shooting, the numbers of the avocet, the white heron in the south, and the pigeons, ducks, swamp hens, quail, and other birds were greatly reduced. In some favoured spots, the slaughter was terrible. In the early days of Otago, the pigeons, which congregated in the fuchsia trees and scrub about Dunedin, suffered severely. A common recipe for soup among the settlers there was: "One kaka parrot and fourteen pigeons." Burning grass, scrub, and bush wrought further havoc, and very soon the quail was almost destroyed, while no crows were left on the eastern side of the South Island. Dogs killed the wood hens in large numbers. Lands were cleared of timber, swamps were drained, the native flora was largely supplanted by crops, hedgerows, and gardens, and by Old World shrubs and trees. The land-birds retreated in diminished numbers, mostly to the mountains, where the ancient forests, still standing, offer them a place of refuge.

In many sanctuaries and forest reserves the birds are holding their own, and some species are increasing in numbers.

THE NEW FAUNA.

New animals were introduced systematically. These greedy invaders soon overcame any feeble resistance that may have been offered, and speedily established themselves.

Some of the new animals were brought in by Europeans for food, and some for purely sentimental reasons. There were few natural food supplies in New Zealand when colonists first came. The Maoris lived mostly on fish, several kinds of birds, fern-root, and three vegetables which they had introduced. It became obvious that if the colony was to be the abode of a large population of civilised beings, acclimatisation must take place on a large scale. Food supplies were, therefore, the first consideration. Sheep and cattle and other animals found a climate resembling that of their native country, and pastures which could not be surpassed. Like their predecessors, the "Captain Cooks," they throve very well and grew fat. Large cattle and sheep runs were formed, and when the needs of the settlers were satisfied, there was established an export trade in wool and mutton that has attained to great dimensions.

Rabbits, pheasants, the honey-bee, and later on, quail, hares, deer, and trout and other food fishes were also introduced. The trout, it may be stated here, feed on the aquatic larvæ of insects that furnish food for birds, and the decline of our two bats may be due to this fact. The white clover did not seed in New Zealand until the honey-bee was imported in 1842, and for a long time the colonists had to import their supplies of red clover. To meet this difficulty, they introduced the humble-bee, and the red clover now seeds freely.

The cultivation of cabbages, cauliflowers, turnips and other succulent plants was followed by an alarming increase in the numbers of native insects. Armies of caterpillars invaded the fields and consumed the crops. It was hardly possible to open a pea-pod without finding a caterpillar inside; and, in the Auckland district, dismayed settlers saw fields of maize under bare poles, not a leaf remaining. The food supply of the insects had been increased enormously, and they were not slow to respond.

It was decided that the best plan to adopt, to make agriculture and horticulture possible, was to introduce insect-eating birds. But it was recognised that these birds must not live on insects alone. There is no winter retreat for insect-eaters in New Zealand, as there is in Europe; and if they could not sustain themselves on vegetable food in the winter months, when the insects were absent, they would perish. The field of selection was therefore restricted to birds which would eat both seeds and insects, which would not try to migrate, and which would become common.

One of the first to be introduced was the sparrow, that unlovely, songless, and impudent vagabond of a bird, denounced by Miss E. A. Ormerod, condemned by the English Board of Agriculture, reviled in America, and outlawed by the Parliament of New Zealand, which called in its help in the time of need. A price is now placed on the sparrow's head,

and poisons and traps are now laid for it, the farmers, the
municipal authorities, and the General Government uniting in
the crusade. There is no doubt that it commits great
depredations on the grain crops, and that it increases
alarmingly. But it must not be forgotten that this rapid
increase is one of the bird's recommendations for the mission
it was introduced to carry out. It certainly checked the
increase of the insects. Without the sparrow, or some other
bird equally common, the dominion would be over-run with the
insects again, and life would be insupportable. No exception
is taken to the means of destruction now adopted, as they
merely lessen the numbers of the sparrows. But it cannot be
admitted that the introduction of this bird was one of the
mistakes in acclimatisation. Those who urge that the sparrow
ought to be banished should name a substitute. Birds
generally eat the food they can obtain with the least trouble.
Although soft-billed birds cannot eat seed, hard-billed members
of the finch family, to which the sparrow belongs, eat insects
as well as seeds. Besides this, the seeds they eat are for the
most part those of weeds; and they destroy the seeds of the
weeds all the year round, while it is only at certain seasons
that they have opportunities for attacking the crops. No one
denies that the seed-eating birds do a great deal of harm as
well as good. Before they are condemned altogether,
however, we should consider whether it is not better to suffer
the ills we have, and mitigate them as much as possible, rather
than revert to others with which we might not be able to
cope.

These statements are made in the face of the almost
unanimous condemnation of the sparrow by the farmers of New
Zealand. In 1906 a vote of the farmers was taken by means
of thousands of circulars, which were circulated broadcast.
When they were returned, it was found that the mass of the
evidence was entirely against the little bird, which was
proclaimed a public nuisance, and was wholeheartedly
condemned.*

*Our Feathered Immigrants, by James Drummond, published by the Government
Printer, Wellington, New Zealand, 1907.

Many other birds were introduced. It is certainly impossible to defend the introduction of some of them, such as the greenfinch, which was liberated in all the provinces, the bullfinch in Nelson, the Java sparrow in Nelson and Auckland and the grass-parrakeet in Canterbury. Fortunately the three last-named failed to establish themselves. In some instances, sentiment clashed with utility. This is so with regard to the skylark. Sentimentally, it is the same "blithe spirit" that delighted Shelley in England, but practically, it is far from being an "embodied joy" to the farmers. We could also have very well done without the blackbird and the house mynah.

The partridge, the turtle-dove, the grey linnet, and the Australian mynah were among the complete failures. The list could easily be added to, but it is only right to say that some birds, such as robin redbreast, did not have a fair chance. It was expected that pheasants would do very well. For a time they increased in both Islands, especially the North. Then they began to decline, and they have now almost died out in many districts, but still are plentiful in parts of the North Island.

The cirl-bunting, the goldfinch, the starling, and perhaps the yellow-bunting spread into Canterbury from Otago. The only cirl-buntings ever turned out in New Zealand were liberated in the Otago Peninsula about 1868. They were not uncommon about Dunedin in 1878, and in 1891 had reached the Malvern Hills, in Canterbury. Goldfinches were abundant in Otago when they were rare or unknown in Canterbury. The same may be said of the starlings, which were comparatively rare in Christchurch in 1880, but were abundant in 1890. Little brown owls (*Athene Noctua*) have established themselves in Otago and Canterbury. Many song-thrushes were liberated in all the provinces of the colony before 1870, but only at Cheviot did they succeed at first. After an interval of ten or twenty years, however, they spread everywhere. The hedge-sparrow has spread to many districts, and has been reported on Campbell Island.

The acclimatisation of deer has been very successful, and a considerable amount of stalking is done in both the North Island and the South. The moose and the wapiti have been let loose on the West Coast of the South Island, and proposals have been made to introduce more big game. The ibex and the thar are also looked upon as likely beasts for the mountainous regions of the South Island; several of the latter have been introduced by the Tourist Department, and have been liberated in the Mount Cook district, in the South Island. The General Government is assisting the Acclimatisation Societies in this direction. It is hoped that the presence of big game will induce sportsmen from other parts of the world to visit the dominion, and will add to its attractions.

On the whole, the results of acclimatisation in New Zealand must be considered favourable, although unfortunate mistakes have been made. Among these the most marked is the introduction of rabbits, about which little need be said, as the disastrous effect of their presence is too well known, and of ferrets, stoats, and weasels.

It is with regret that we have to pass by the scientific side of this subject. Naturalists in all parts of the world would read with keen interest a detailed account of how each kind of animal behaved when it was first turned out here. They would like to know how it adapted itself to its new conditions, or why it failed to do so and perished. Why has the goldfinch become ubiquitous, and, together with the blackbird, spread unassisted to the Auckland Islands, while other English birds are confined to small districts? It would be extremely interesting to explain these things. Such explanation as is possible, however, must be left to enthusiastic field naturalists, who are willing to watch long and closely, and who may still clear away some of the difficulties. An attempt has been made in *Our Feathered Immigrants* to deal with the subject on a comprehensive scale. The publication was issued by the Agricultural Department, which gave the author much assistance in collecting the material. All the naturalised English birds are dealt with in the publication. A list of dates on which introduced birds were liberated or appeared in different districts of New Zealand, is published in the *Transactions of the New Zealand Institute*, Vol. xxxix. (1906), p. 503. The Hon. G. M. Thomson, M.L.C., Dunedin, has published *The Naturalisation of Animals and Plants in New Zealand*, dealing with forty-eight species of mammals, 130 species of birds, and five species of amphibians introduced into New Zealand.

The nomenclature of the birds of New Zealand has been exhaustively revised by Messrs. G. M. Mathews and T. Iredale, in a Reference List in *The Ibis* (London). Part I., April, 1913, p. 201; Part II., July, 1913, p. 402.

MAMMALIA

Body usually covered with hair; the young fed with milk from the mother.

ORDER CHIROPTERA.

Fore limb modified into a flying organ by the elongation of the digits.

Family Emballonuridae.

Nostrils simple, ears large, the tragi minute. First joint of the middle finger, when in repose, folded on the wrist. Tail partially free from the inter-femoral membrane. Temperate and tropical regions of both hemispheres.

Genus Mystacops.

Ears separate, the tragus long and attenuated. Middle finger of three joints. Legs short. Tail perforating the inter-femoral membrane, and appearing on its upper surface. New Zealand only.

The Short-tailed Bat.

Mystacops tuberculatus.

Above brown, paler below; long erect hairs fringe the lips. Length of the third finger 3 in. Both Islands. One large colony has been found on the Little Barrier Island, and individual members of the species are reported from Waikawa, in Southland. The presence of an additional joint in the middle finger of this bat enables the folded wings to occupy a very small space; the first joint of the third finger is bent upwards in repose, and not downwards, under the wrist, as in the rest of the family; and the wing, when folded, is thus tucked in beneath the thickened portion of the membrane; while, at the same time, the posterior half of the inter-femoral membrane, from the point where the tail perforates it, is rolled upwards and forwards beneath the leathery anterior half. The fur is very peculiar. The hairs are moderately long, and much thicker than in any other species. Viewed under a microscope, the shaft of each of the long hairs appears almost smooth, with very slight indications of the margins of the hair-scales so conspicuous in every other kind of bat.

The presence of two bats saves New Zealand from being styled a country without land mammals. Although the Maori dog and the native rat were once catalogued among the dominion's mammalia, it is now generally agreed that these two animals were brought by the Maoris at the time of the great migration from Hawaiki, the mother country New Zealand has no more right to claim them as its own than it has to claim Clydesdale horses or Southdown sheep.

As the dog and the rat were the only candidates, except the seals, for a division of the honours, the bats are left in undisputed

Short-tailed Bat.   *(Voy. Erebus and Terror.)*

possession, and are recognised as practically the sole land representatives of this dominion in the highest class of the Animal Kingdom. Our poverty in regard even to bats has caused comment, and Dr. Wallace, in his *Island Life*, says that there is a very remarkable contrast between New Zealand and the British Isles, where there are at least twelve distinct species, though the climate there is far less congenial to animal life.

New Zealand's bats are popularly called the short-tailed and the long-tailed. As if to make up in one respect for deficiency in another, short-tail has long ears, and long-tail has short ones.

The short-tailed species is remarkable and rare. To naturalists it is one of the most interesting bats in the world. So marked are its peculiarities, that it has been placed in a separate genus. There are at least eighty genera known to science, and there is an enormous number of species; but New Zealand's short-tailed bat occupies a unique position, and is apart and distinct from them all. It belongs to no other country, and has been seen so seldom that very little is known about its habits of life. Some of these, however, may be guessed at, as they are suggested by the animal's physical peculiarities, which have attracted special attention. Most bats are well adapted for life in the air, feeding on flying insects, and even drinking on the wing. But our short-tailed species has adaptations which lead to the conclusion that it hunts for its insect food not only in the air, but also on the branches and leaves of trees, among which its peculiarities of structure must enable it to creep and crawl with ease and security In cold climates, as winter approaches, bats seek shelter in caverns, vaults, and ruined and deserted buildings, where they remain in a torpid state until returning spring calls them to active exertions. But in a sub-tropical climate, from which this creature's ancestors came, bats are not likely to hibernate, and as there are no flying insects in New Zealand in the winter months, the climbing habit may be necessary for this little bat's existence.

Family Vespertilionidae.

Nostrils simple. Ears moderate, with large tragi. The first joint of the middle finger, when in repose, extended in a line with the wrist. Tail contained in the inter-femoral membrane. Temperate and tropical regions of both hemispheres.

Genus Chalinolobus.

Muzzle short and obtuse. Ears short, the tragus expanded above and turned inwards. Lower lip with distinct fleshy lobule near the angle of the mouth. Australian and Ethiopian regions.

The Long-tailed Bat.—Pekapeka.

Chalinolobus morio.

Dark brown on the head and neck, passing into dark chestnut brown posteriorly. Length of the third finger 2.7 in. New Zealand and south-east of Australia.

The long-tailed bat, with short ears, is spread over New Zealand and is much more numerous than the short-tailed bat. It belongs to a species that is found in New Zealand and south-east Australia, to a genus that occurs in Australian and Ethiopian regions, and to a family that is cosmopolitan. Up to about 1885,

Long-tailed Bat.   *(Voy. Erebus and Terror.)*

it was common about Christchurch, but it is thought that the destruction of the old wooden bridges over the Avon, where numbers used to gather together, has driven it away It measures about two inches in total length, being slightly smaller than the other species, and is about the same size as the "flittermouse," the commonest species in England.

At one time it was thought that the long-tailed bats, unlike those in the Old Country, lived solitary lives, and were averse from forming themselves into communities. Experiments and closer examinations, however, have proved that great numbers of our bats live together in hollow trees in the bush. Mr. R. Caldwell, a district surveyor in the North Island, supplied Sir

Walter Buller with some interesting information in this respect. "I left Carterton, together with two companions," he says, "for a walk into the hills, at the right-hand side of Waiohine, going by way of the Belvedere Road. We got fairly up the hills by about 10 a.m., and climbed a high range, covered with black beech. Getting warm, we sat down on the moss to rest. Then my attention was attracted by a smell of a kind I had not noticed in the bush before, and one that reminded me of a flying-fox camp in Queensland. I followed the smell for some distance to a large beech tree, with an opening about four feet from the ground. I had evidently traced the smell to its source, for at the opening it was fairly stifling. I could see nothing, so I lighted a bunch of dry leaves and thrust it through the opening into the tree. As I did this, a bat flew out in my face, another, and another. The smoke increased, and the bats streamed out in hundreds. I had no means of computing the number; but one of my men, having a small switch in his hand, kept striking at the stream, the result of which I afterwards counted. There were exactly a hundred bats killed. For one killed, at least ten must have passed and flown away Large numbers dropped down in clusters through the blazing opening. I had no idea there were so many bats in the Wairarapa, and would not have believed it had I not seen them. I have never seen in New Zealand such another collection."

In 1893, a man took twenty-two living bats in a box to Mr. T. F. Cheeseman's office in Auckland. He stated that he had been bush-felling near Reweti, on the Kaipara railway. A tree was cut down, and as it struck the ground, the men were surprised to see numbers of bats fly from the upper branches. Running to the spot, they found clusters of the creatures still clinging to the branches, and they collected about thirty.

Mr. Cheeseman, being anxious to see how they would behave in a room with closed doors and windows, liberated them. The experiment justified, to some extent, the belief that bats enjoy an acute sense of touch, probably unequalled throughout the animal kingdom. They took to their wings at once, and commenced to circle round the room with that quick, soft, and noiseless flight which they are enabled to pursue by means of their velevty wings. The presence of full daylight did not affect them in the slightest degree, and they made no mistake in estimating their distance from an object. They circled round the room, flying in and out of the corners, skimming just below the ceiling, and hovering over the furniture, but never coming in contact with anything. Nor did they dash themselves against the window-panes, as birds would have done in similar circumstances, but they treated the glass in precisely the same manner as the walls of the room. After satisfying themselves that there was no mode of escape from the room, they began to settle down on the tops of the architraves of the doors and windows, hanging head-downwards, by the claws of their hind-wings. Ultimately, they collected in clusters of four or five, cuddling quite close to one another, and they were then quite easily transferred to their cage.

ORDER CARNIVORA.

Flesh-eating mammals, with sharp teeth.

Section Pinnipedia.

Legs modified into flippers.

Key to the Families.

Without external ears.	Phocidæ
With external ears.	Otariidæ

Family Otariidae.

Ears small. Hind limbs capable of being turned forwards. Fore flippers with rudimentary claws. Incisors, six above and four below.

Genus Arctocephalus.

Muzzle rather tapering in front, ears rather long. Palate of the skull rather narrower behind than in front; short, scarcely reaching to the middle of the zygomatic arch. Southern Seas.

Key to the Species.

No under-fur mixed with the hair.	A. hookeri.
Under-fur as well as hair.	A. forsteri.

The Sea Lion.—WHAKAHA.

Arctocephalus hookeri.

Male—Greyish black above, dark brown below; the neck very thick, and with longer hairs. Length, when full grown, ten or twelve feet. Female—Pale tawny. Length, six feet. Young—At first like the female, then brownish black. Auckland Islands and Campbell Island. The male only is called sea lion by the sealers, the female is generally called sea bear. It is still very common at the Auckland Islands.

The Auckland Islands were the scene of sealing activity for many years, the first expeditions setting out from Sydney early

(Trans. Zool. Soc.)

Sea Lion.

in the nineteenth century. The industry, which was conducted on a large scale, reached the height of its prosperity between 1810 and 1820. The sealers had to suffer great hardships, but they were attracted by the lucrative trade, and large numbers of seals were slaughtered in the race for wealth.

Captain T. Musgrave, in his interesting journal entitled *Cast Away on the Auckland Islands*, published in 1866, gives detailed accounts of the seals' habits. Dealing with the sea lion, he says:—"The females are of a grey, golden buff, or beautiful silver colour, sometimes spotted like the leopard, and they are called tiger seals. Their fur is about an inch long, not very soft, but very thick, and particularly sleek and smooth. The males are uniformly of a blackish grey colour. The fur and skin of the male are superior to those of the female, being much thicker. On the neck and shoulders the male has a thicker, longer, and much coarser coat of fur, which may almost be termed bristles. It is from three to four inches long, and can be ruffed up and made to stand erect at will.''

In regard to the movements of the animals on land, Captain Musgrave says:—'' They go roaring about the woods like wild cattle. When they are on shore, they can run surprisingly fast; on the hard, smooth beach they can run nearly as fast as a man, and in the bush, or in the long grass, faster than a man. They are able to climb up rocky cliffs and steep slippery banks that would be inaccessible to men. The bulls are very bold, and will come out of the water and chase us. They are particularly fierce.''

He adds that the bull seals go into the bays in October or the beginning of November, when they are very fat. During those months, and the two following ones, seals pass most of their time on shore, the bulls basking in the sun, while the cows roam about the woods looking for a suitable place in which to calve. The young are born early in February, always on shore, and there is only one calf at birth. When the young are a few days old, the mother sets to work to get them into the water, but they show great antipathy to it, and only after many efforts, and much biting, beating, and pushing, does the mother succeed. One mother has been known to spend over three days in getting her calf half-a-mile towards the water. After this, the cows assemble in mobs of from twelve to twenty, with their young, and pass the months of February, March, and April, when they suckle the young, chiefly on shore. There are generally one or two bulls in each mob, but these leave the bays after the beginning of April. When the young are three months old, they leave off suckling, and, with their mothers, keep the water during the day, but return to shore at night to sleep, going to the water again before daylight. They do not choose any particular place for sleeping in, but take shelter anywhere in the bush, or in the long grass close to the shore. If not disturbed, a mob will sleep in the same place many nights, but, if they are disturbed, they will shift their camp.

It was generally supposed, up till 1903, that our seals migrated in the same manner as those of the north; but Dr. Cockayne, who visited the islands in the winter of that year, reports having seen sea lions in large numbers. It seems, therefore, that some do not migrate, but remain at their homes all the year round.

The Fur Seal.—KEKENO.

Arctocephalus forsteri.

Blackish-brown, or grizzled when old, and with reddish chestnut under-fur (principally on the back), which is white near the base. The sexes are nearly alike in size and colour. Length of the male, 5½ to 7ft.; of the female, 4½ to 7ft. New Zealand and the Southern Islands.

In the early days of Australian colonisation, sealing was carried on in Bass Strait, and, when the industry began to flag there, attention was turned to the coast of New Zealand, and to the neighbouring islands. It was not long before a regular trade in connection with the fur seal was established with Sydney. But the slaughter of the seals was so great that, after a few years had passed, they became scarce, and the industry came to an end.

Captain Cook, in his account of his first voyage, mentions the fact that seals had been seen by him in New Zealand and only on the coast of the South Island. On his second voyage, the great navigator recruited his ship's crew at Dusky Bay, and the first fresh meat eaten by him in that locality was obtained from the first seal killed in Dusky. During his stay of nearly two months at the Southern Sounds, Cook killed many seals, using the flesh for food, the skins for repairs to the rigging, and the blubber for oil for his lamps.

The late Hon. R. McNab, a former Minister for Lands, in a communication to the authors, supplied some very interesting information on this subject. He stated that in 1792 the first sealing gang was landed on the shores of New Zealand, and was stationed at Dusky for nearly a year, procuring 4500 skins for the employer of the gang, Captain Raven, of the *Britannia*. In 1803, when the Bass Strait sealing began to decline, some of the boats determined to try further fields, and Messrs. Kable and Underwood, of Sydney, commenced the trade between Sydney and New Zealand with a schooner of 31 tons, called the *Endeavour*. She visited Dusky and Breaksea Sounds and the Solander Island, and received a cargo of 2000 skins, which realised 4s. 6d. each. From that time onward the trade flourished. Sir Joseph Banks, writing in 1806 about seals in the southern parts of New Zealand, says: " Their stations on the rocks or in the bays have remained unmolested since the Creation. The beach is encumbered with their quantities, and those who visit their haunts have less trouble in killing them than the servants of the victualling office have, who kill hogs in a pen with a mallet." In July of the same year, Sir Joseph stated that on one vessel expected from Sydney there were 30,000 sealskins from the Antipodes Islands.

Mr. McNab adds that as early as 1810 the effect of the reckless slaughter began to be felt, and it was the discovery of the Campbell Islands and the Macquarie Islands, in that year, which gave fresh life to this trade. The number of seals taken from the southern islands was enormous. After 1810, however, only those captains who were well acquainted with the seals' haunts were able to obtain large quantities of skins.

It has been calculated that, in 1824, ten vessels touched on the New Zealand coast and at the islands, and took away 70,000 or 80,000 skins, mostly from the southern islands, 40,000 or 50,000 going to Sydney, and the remainder to England. In those days, a sealskin in Sydney was worth about 15s. It is recorded that the industry was carried on so assiduously, and the south-west portions of the coast were hunted so industriously by sealers, who killed the females and the young for food, that there was fear of the seal, in those parts, becoming extinct. Two years later, a vessel spent six months in a cruise searching for new sealing grounds, and took back only 449 skins. Stewart Island was a specially favoured spot for sealers, and many of them also established stations in Dusky Sound and in the bays and inlets on the coast. The fur seal is still found at the Snares, the Bounty Islands, the West Coast Sounds, and the seal rocks off Westport.

Family Phocidae.

Hind limbs projecting backwards, and not capable of being turned forward. Fore flippers smaller than the hind ones. Fore flippers with claws, hind flippers often without them. No external ear.

Key to the Genera.

Four incisors in the lower jaw.	Ogmorhinus.
Two incisors in the lower jaw.	Macrorhinus.

Genus Ogmorhinus.

All the molars, except the first, with two roots. Molars with three pointed cusps, of which the middle one is the longest. Antarctic Seas.

(*Southern Cross Coll. Brit. Mus.*)
Sea Leopard.

The Sea Leopard.—Pakaka.

Ogmorhinus Leptonyx.

Grey, paler beneath, with small black spots on the neck and body, and some white spots on the sides. Usual length 7 to 8 feet, but sometimes 12 feet. An occasional visitor to the shores of New Zealand.

The natural home of the sea leopard is in the ice-pack and the islands near it; but it sometimes wanders north to New Zealand, Tasmania, and New South Wales. It gives birth to its young in September. It feeds on fish, birds, especially penguins, and brown sea-weed. It is a savage animal, and will readily attack man, but very little is known of its habits. It weighs about 850 lbs.

Genus Macrorhinus.

Molars small, with a single root and plaited crown. The canines large. Muzzle hairy; male with an exsertile proboscis. Antarctic Seas.

The Sea Elephant.

Macrorhinus leoninus.

Uniformly brown. Length of the male, about 18 feet; of the female, about 10 feet.

Sea Elephants.   *(Drawing by C. E. Bickerton.)*

This is the only species. It is found at Macquarie Island and Campbell Islands as well as other islands in the Antarctic Seas. It is killed for its blubber, not for its fur.

ORDER CETACEA.

Body fish-like, with a tail-fin composed of a pair of horizontal flukes.

Key to the Families.

1. Mouth with baleen.		Balænidæ
Mouth without baleen.		2
2. Teeth in both jaws.		Delphinidæ
Teeth in lower jaw only.		3
3. Teeth in lower jaw numerous.		Physeteridæ
Teeth in lower jaw, one or two pairs.		Ziphiidæ

Family Balaenidae.

Teeth never developed, but the palate with plates of baleen, or whalebone. Skull symmetrical.

The whalebone whales form the most important group, and include the right whale, the hump-back, the fin-back, or rorqual, and other species less notable. Their teeth are never developed. The family derives its name from baleen, or whalebone, attached to the upper jaw or palate. It acts as a sieve to strain off the water taken into the mouth when the whale is feeding on the small fry of the ocean. Sir W H. Flower states that the immense mouth is filled with water containing shoals of the small creatures, and, on the whale closing its jaws and raising its tongue so as to diminish the cavity of the mouth, the water streams out through the narrow intervals between the hairy fringe of the whalebone plates, and escapes through the lips, leaving the living prey to be swallowed. Each plate of baleen, of the northern right whale, weighs about seven pounds. From a large whale there may be obtained about a ton and three-quarters. Before being used, the whalebone is made soft and pliable by being immersed in boiling water for about twelve hours.

Key to the Genera.

1. Throat smooth, baleen long.	2
Throat plaited, baleen short.	3
2. No dorsal fin, baleen black.	Balæna.
A dorsal fin, baleen yellow with a black edge	Neobalæna.
3. Pectoral fin long.	Megaptera.
Pectoral fin short.	Balænoptera.

Genus Balaena.

Head about two-sevenths of the total length. Skin of throat smooth. No dorsal fin. Mouth much arched. Length of the baleen six or seven times the breadth at the base.

The Southern Right Whale.—Tohora.

Balaena australis.

Entirely black. Length up to 70 feet; that of the baleen, 6 feet.

The southern right whale was once very abundant, the Tasman Sea and the Pacific Ocean between the Chatham Islands and Norfolk Island being the chief cruising grounds for whalers. The females visited the shallow bays to calve in May, June, and

July, and were joined there by the males in August or September, and all went to sea together.

In the early days of New Zealand, whales frequented these waters in very large numbers, and an extensive whaling industry was carried on. At one time, about 300 vessels, chiefly from America, visited the colony every year. The industry began about 1795, reached the height of its prosperity between 1830 and 1840, and then dwindled away. In 1835, 116 vessels called in at the Bay of Islands alone, and many stations were established along the coast of the mainland, and at Kapiti and Stewart Islands; but whales seldom visit New Zealand waters now, and the industry has almost disappeared with them.

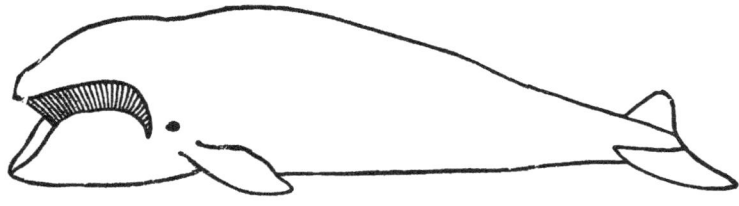

Southern Right Whale.

Between 1830 and 1840, the Island of Kapiti, and the other small islands in the vicinity, were largely patronised by whalers. Dr. Dieffenbach states that, in 1839, the produce of the establishments on these islands was 466 tuns of oil, and 30 tuns of whalebone, obtained by twenty-three boats.

Professor Beddard, in his work entitled *A Book of Whales*, published in 1903, says that a very singular feature of Balæna is the horny, irregular mass growing on the snout, called the "bonnet." The irregular shape and pitted appearance of the bonnet, he says, gives the impression that it is a pathological structure, a kind of corn, perhaps produced by the animal rubbing itself against rocks, as this species has been observed to do in order to get rid of the barnacles which are apt to infest it. Captain Scammon states that towards the end of the season these whales congregate in herds, which are technically known as "gams." The right whale is a slow swimmer, making not more than four miles an hour. It feeds on small animals floating at or near the surface of the sea.

Genus Neobalaena.

Head about one-fourth of the total length. A small falcate dorsal fin.

Australian Whale.

Neobalaena marginata.

Baleen yellowish white, with a narrow back margin. Length of the whole animal, 20 feet. Of the baleen, 2 feet. New Zealand and Australia. Found on all parts of the coast, but said by the whalers to be rare.

Genus Megaptera.

Head moderate. Skin of the throat plicated. The dorsal fin very low. Flippers long and narrow, about one-fourth of the length of the animal.

The Hump-back.—PAKAKE.

Megaptera boops.

Black, except the flippers, which are white. Baleen black. Usual length, 45 to 50 feet.

The hump-back is known by its long flippers, which are indented along the margins. This whale is solitary in habit, and the baleen is short and coarse. The hump-back is found in many seas. According to Captain Scammon, it is irregular in its movements, seldom going in a straight course for any considerable distance; at one time, he says, these whales may be seen moving about in large numbers, scattered over the sea as far as the eye can reach from the masthead, at other times singly.

Genus Balaenoptera.

Head small and flat. Skin of the throat plicated. A small falcate dorsal fin. Flipper small and pointed. Baleen short and coarse. Length of the baleen, three or four times the breadth at the base.

Key to the Species.

Baleen black.	B. Sibbaldi.
Baleen slate-blue.	B. musculus.
Baleen yellowish white.	B. rostrata.

The Blue Whale.

Balaenoptera sibbaldi.

Dark bluish grey, with small whitish spots on the breast. Flippers about one-seventh of the total length. Baleen black. Total length, up to 87 feet; of the baleen, 2 feet; breadth at the base, 1½ feet.

Professor Beddard states that this whale, which was named in honour of Sir R. Sibbald, may be distinguished from others by its superior size, and by various other characters. The whalers know it by its largeness, and by the height to which it spouts; and its speed, he adds, is sometimes great, as it can accomplish about twelve miles an hour when at its best.

The largest specimen recorded up to the present time (1922) was stranded on the beach at Okarito, on the West Coast of the South Island, in the middle of 1908. It was 87 feet long and 16 feet high. It was purchased by Mr. Edgar F. Stead, of Christchurch, and Mr. R. Turnbull, of Wellington, and the skeleton was sold by them to the authorities of Canterbury Museum, Christchurch, for £500, of which £200 was subscribed by the public of Canterbury, and Messrs. Stead and Turnbull contributed £50 each to the funds.

The Fin-back or Rorqual.—RATAHUIHUI.

Balaenoptera musculus.

Greyish slate-colour above, and white below. Baleen slate-colour, vareigated with yellow or brown. Length of the animal 65 to 70 feet.

The fin-back is widely distributed, being found even in tropical regions. It is very active, and does not yield much blubber. Its principal food is fish. When struck by a harpoon, it displays greater fierceness and boldness in its movements, it is stated, than do other species of whales. It has been known to turn suddenly against the whaling boats and dash them to pieces by strokes of its tail. It is said that more than 5000 plates of baleen, or whalebone, are contained in the mouth of the rorqual; but the plates are too short and too coarse to be valuable. It is estimated that the blubber is only about six or eight inches in thickness as a rule, and that only ten or fifteen tons of oil can be obtained from an average specimen. Its ordinary rate of speed is about five miles an hour.

The Pike Whale.

The Balaenoptera rostrata.

Greyish black above and white below. Baleen yellowish white. A white band across the flippers. Length, 30 feet.

This whale, according to Professor Beddard, pursues fishes, and dog-fishes have been discovered in its stomach. It has also been stated that the stomach contains pebbles. "This is curious,"

Pike Whale.

(Ann. Mag. Nat. His.)

Professor Beddard says, "for, in other whales, and in sea lions, the same observation has been made, and possibly in both cases the stones were taken up accidentally while in pursuit of fish: one can hardly believe that any idea of ballast entered into the mind of the cetacean."

Family Physeteridae.

No functional teeth in the upper jaw. Skull asymmetrical about the blow-holes.

Key to the Genera.

Large—20 to 25 pairs of teeth.	Physeter.
Small—9 to 12 pairs of teeth.	Cogia.

Genus Physeter.

Teeth stout and conical. Head about one-third of the length of the body, very high and truncated in front. Blow-hole single, at the anterior end of the head. Flippers short and broad. Dorsal fin very low.

The Sperm Whale.—PARAOA.

Physeter macrocephalus.

Black above and grey below.   Length, from 55 to 60 feet.

The sperm whale frequents tropical and sub-tropical seas, although it was formerly very common off Stewart Island. It feeds on both fish and cuttle-fish. It is one of the largest of the cetaceans, its length being generally from 55 to 60 feet. It is an ungainly creature, but is very valuable, chiefly on account of the

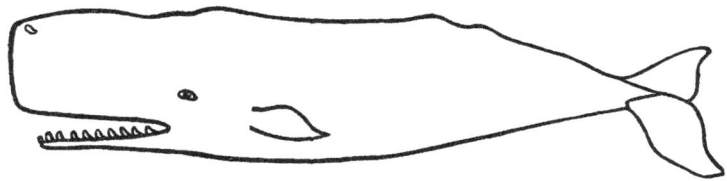

The Sperm Whale.

spermaceti, which is contained within the head in two large special cavities, and also diffused through the blubber. These whales are gregarious, and go about in "schools," each "school" being generally led by a few old males. Sperm whales have been known to turn with great determination upon boats that were pursuing them, and Captain Scammon states that they have attacked even ships. According to Professor Beddard, the cachalot, as it is sometimes called, will remain under water from fifty minutes to an hour and a quarter. "When it spouts, it does so for the space of about three seconds, and the column of water ejected can be seen from the masthead at a distance of three to five miles. The spouting of the sperm whale can be readily distinguished from that of other whales owing to the fact that the blowhole is single, and the column of breath condensed is also a single fountain, not a double jet, as in other whales. Moreover, as the blowhole is situated further forwards, than in other whales, the jet is not directed upwards, but forwards." The sperm whale swims at the rate of about ten miles an hour.

Genus Cogia.

Teeth in the lower jaw, 9 to 12 on each side; slender, pointed, and curved.

The Pigmy Whale.

Cogia breviceps.

Black above, greyish white below. Length, 8 to 10 feet.

Family Ziphiidae.

Teeth of the lower jaw rudimentary, except one or two pairs. Head beaked; the blowhole single, median. Skull asymmetrical about the blowholes.

Key to the Genera.

Two pairs of teeth at anterior end	Berardius.
One pair of teeth at the anterior end.	Ziphius.
One pair of teeth near the middle.	Mesoplodon.

Genus Berardius.

Two moderate-sized teeth near the end of the lower jaw. Beak about half the length to the eye. Blowhole over the eye. Dorsal fin rounded at the end. New Zealand only.

Porpoise Whale.

The Porpoise Whale.

Berardius arnuxi.

Black, with a narrow band of white along the belly. Length, up to 30 feet.

Genus Ziphius.

A single conical tooth on each side of the lower jaw near the end. Beak very small, gradually tapering. Dorsal fin rounded at the apex.

The Goose-beak Whale.

Ziphius cavirostris.

Black, varied with white, sometimes on the lower surface, sometimes on the upper. Length, 15 to 20 feet.

The goose-beak is a rare and solitary animal, found in the temperate regions of both hemispheres. It feeds on cuttle-fish.

(Trans. Zool. Soc., Vol. XII.)

Goose-beak Whale.

The Scamperdown Whale.—HAKURA

Genus Mesoplodon.

A much compressed and pointed tooth on each side in the lower jaw at some distance behind the apex. Dorsal fin pointed at the apex. Beak long. Colour black, variously mottled with white, especially on the top of the head. Length, up to 20 feet. The species can be distinguished only by the skulls.

Scamperdown Whale.

Key to the Species

1. Teeth near the apex of the jaw.   M. hectori.
   Teeth near the middle of the jaw.   2
2. Both pairs of foramina near the base of the rostrum, on
   the same level.   M. layardi.
   The larger pair of foramina near the base of the rostrum.
   in front of the smaller.   3

3. Rostrum higher than broad.   M. haasti.
   Rostrum broader than high.   4
4. Rostral groove deep.   M. australis.
   Rostral groove shallow.   M. grayi.

Family Delphinidae.

Teeth usually numerous in both jaws. Blow hole transverse. crescentic, with the horns of the crescent pointing forward.

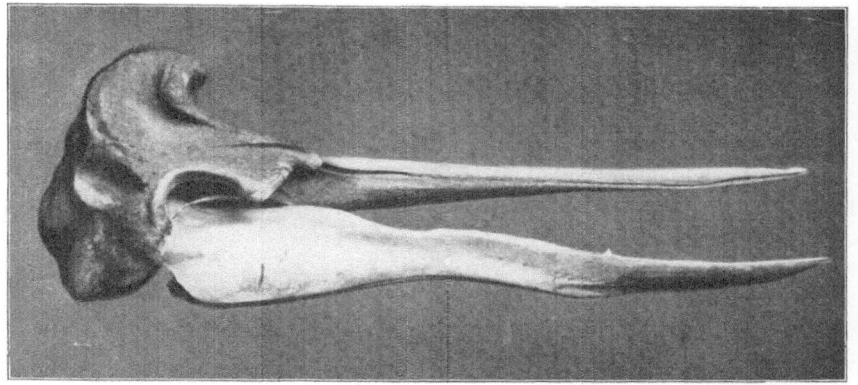

(Photo. Burton Bros.)

Skull of Scamperdown Whale.

Key to the Genera.

1. Not beaked.   2
   Head beaked.   5
2. Teeth more than 20 in each jaw.   Cephalorhynchus.
   Teeth less than 20 in each jaw.   3
3. Teeth very strong.   Orca.
   Teeth small.   4
4. A dorsal fin.   Globicephalus.
   No dorsal fin.   Delphinapterus.
5. Teeth less than 30 in each jaw.   Tursiops.
   Teeth 30 or more in each jaw.   6
6. Palate of skull grooved.   Delphinus.
   Palate of skull not grooved.   Prodelphinus.

Genus Delphinapterus.

Teeth, from eight to ten on each side of the jaw, rather small, separated by intervals considerably wider than the diameter of the teeth. Flippers short and broad. No dorsal fin, but a low ridge in its place.

Beluga, or White Whale.

Delphinapterus leucas.

White.   About 12ft. long.

The white whale is easily tamed, and is very docile.   Its blubber yields considerable quantities of oil.   It is seldom seen

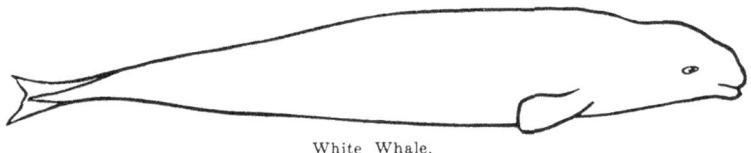

White Whale.

in these waters.   In Hudson Bay and Davis Straits, the white whales, which are there found in large numbers, are hunted for their skins as well as their oil.   The Greenlanders depend largely for sustenance on this whale.

Porpoise.
(*Drawing by A. W. Walsh.*)

Genus Cephalorhynchus.

Teeth small, from 25 to 30 each side of the jaw.   Flippers rather small.

The Porpoise.—Upokohue.*

Cephalorhynchus hectori.

Dorsal fin flatly rounded at the apex; flippers slightly longer than the distance from the muzzle to the eye. Teeth, 31 or 32 in each side of the jaw. Above, pale grey. Lower jaw, throat, and belly white, the white on the belly being divided by a transverse band of grey just behind the flippers; nose and forehead white; a white band from below the dorsal, sloping obliquely upward and backward towards the tail. Sides of the head, a transverse band just behind the blowhole, and the flippers, dark slate grey. Length, 4 to 5 feet. Teeth, five in an inch. Abundant round the coast of New Zealand.

The porpoise is gregarious and frequents the coasts, never going far from land. It feeds on fish.

Dolphin.

Cowfish.

(*Trans. Zool. Soc., Vol. XI.*)

Genus Delphinus.

Teeth very numerous in both jaws, small, close set. Rostrum, elongated, the palate with deep lateral grooves.

---

*Archdeacon Williams states that he knows the names Tupoupou and Waiaua; he has Upokohue for "black-fish."

The Dolphin.

Delphinus delphis.

Beak at least half the length of the gape. Flippers shorter than the gape. Teeth about 45 in each jaw. Above brown; below white, the white extending over the eye. Fins dark, the dorsal and pectorals often with a large white blotch. Length about six feet. The North Island as far as Cook Strait.

The dolphin chiefly frequents the open sea, often far from land. It is gregarious, and feeds largely on flying-fish.

Genus Orca.

Teeth, about 12 on each side of the jaw, very large and stout. Flippers very large, nearly as broad as long. Dorsal fin near the middle of the back, very high and pointed.

The Killer Whale.

Orca gladiator.

Dark above, and white below; a white mark over the eye.

The killer whale has been described as the wolf of the ocean. Several of these whales, which are active and ferocious, unite in attacking the larger species. Mr. F. Bullen was once an onlooker at a terrible fight in the waters, in which three killers were the aggressors. He says: ''The first inkling of what was really going on was the leaping of a killer high into the air by the side of the whale, descending upon the victim's broad smooth back with a resounding crash. I saw that the killer was provided with a pair of huge fins, one on his back, the other on his belly, which at first sight looked as if they were also weapons of offence. A little observation convinced me that they were fins only. Again and again the aggressor leapt into the air, falling each time on to the whale's back, as if to beat him to submission. The sea foamed and boiled like a cauldron, so that it was only occasional glimpses I was able to get of the killers, until presently the worried whale lifted his head clear out of the surrounding smother, revealing two furies hanging, one on each side, to his lips, as if endeavouring to drag his mouth open, which I after-wards saw was their principal object, as whenever, during the tumult, I caught sight of them, they were still in the same position. At last the tremendous and incessant blows, dealt by the most active member of the trio, seemed actually to have exhausted the immense vitality of the great 'bowhead,' for he lay supine upon the surface. Then the three joined their forces, and succeeded in dragging open his cavernous mouth, into which they freely entered, devouring his tongue. This, then, had been their sole object, for as soon as they had finished their barbarous feast, they departed, leaving him helpless and dying, to fall an easy prey to our returning boats.''

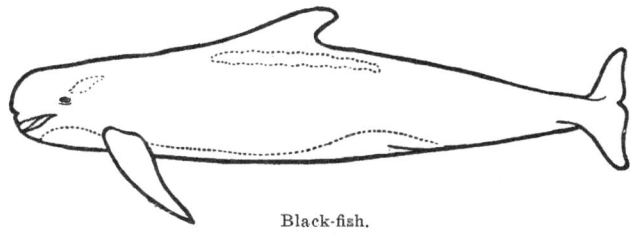

Black-fish.

Genus Globicephalus.

Teeth, 8 to 12, confined to the anterior half of the jaw; small and conical, and widely separated. Flippers very long and narrow. Fore part of the head rounded. Dorsal fin low and triangular, the length of the base more than the vertical height.

The Black-fish.

Globicephalus melas.

Nearly uniform black in colour, except in the middle of the under surface, which is lighter. Length, 20 feet.

Gregarious and timid in disposition. Feeds on cuttle-fish.

Genus Tursiops.

Teeth, 21 to 25 in each jaw, stout. Beak tapering slightly from the base to the apex. Flippers moderate, narrow.

The Cow-fish.

Tursiops tursio.

Beak rather less than one-third of the gape. Flippers equal to the distance between the muzzle and the eye. Teeth, about 22 in each jaw, two or three in an inch. Above, including the upper jaw, dark slate-blue, passing gradually into white below. Fins, slate-blue. Length seven to ten feet. Southern parts of New Zealand.

Genus Prodelphinus.

Teeth, 30 to 50 in each jaw, small. Plate of the skull without lateral grooves.

The Bottle-nose.

Prodelphinus obscurus.

Beak short, but very distinct. Dorsal fin falcate. Flippers longer than the distance from the muzzle to the eye. Teeth, 24 to 28 in each jaw, about five to an inch. Back and fins blackish, the muzzle and belly white. A white band from below the dorsal fin sloping obliquely downward and backward towards the tail. Length, about five feet. Northern parts of New Zealand. Common in Cook Strait.

Genus Grampus.

Risso's Dolphin.

Grampus griseus.

No teeth in upper jaw, but three to seven on each side of mandible near to the symphysis. No beak. Pectoral fin long, pointed, falcate; dorsal fin high and falcate. Length from 10 to 13 feet. Grey, varying on the fins and tail to black, and to white on the belly. Mainly Mediterranean and North Atlantic in range. There is only one species in the genus. (Beddard.)

As stated in the Introduction, " Pelorus Jack " is officially gazetted as a Grampus dolphin. It is the only member of its genus recorded in New Zealand waters. Professor Beddard states that the Grampus, like many other cetaceans, has no very fixed limits to its range, and it may go from the Mediterranean and North Atlantic to southern latitudes during the winter. He adds that it is not by any means a common cetacean, and that only a dozen records of its capture on the English and French coasts are extant. It feeds on cuttle-fish. Up to the present time, April, 1923, is has not been reported since November, 1916.

(*Photo. by A. Pitt.*)

"Pelorus Jack."

AVES

SUB-CLASS CARINATAE.

Sternum with a keel. Feathers with hooklets, the webs firm.

Key to the Orders.

This key is intended to help beginners in naming a bird by directing them to the order, or sub-order, to which it belongs. To use it, begin at No. 1 on the left, and then follow to the number on the descriptive line, answering to the specimen under examination.

1. Toes without a membrane.	2
Toes united by a membrane.	12
Toes margined on each side.	Pygopodes.
2. Wings minute, hidden.	Apterygiformes.
Wings exposed.	3
3. Toes three before and one or more behind.	4
Toes two before and two behind.	16
4. Bill hooked, claws very sharp.	Raptores.
Bill not hooked, claws moderate or blunt.	5
5. Inner toe partly joined to the middle.	Halcyones.
Inner toe free.	6
6. Lower portion of tibiæ feathered.	7
Lower portion of tibiæ naked.	10
7. Bill soft and swollen at the base.	Columbiformes.
Bill hard, not swollen at the base.	8
8. Hind toe at same level as the others.	Passeres.
Hind toe elevated above the others.	9
9. Bill long.	Ralliformes.
Bill short.	Galliformes.
10. Hind toe at same level as the others.	Herodiones.
Hind toe elevated or absent.	11
11. Bill stout, compressed.	Ralliformes.
Bill slender.	Limicolæ.
12. The front toes only connected.	13
All four toes connected.	Steganopodes.
13. Wings long and pointed.	14
Wings with rudimentary feathers.	Impennes.
14. Nostrils in tubes, bill hooked.	Tubinares.
Nostrils exposed, bill not hooked.	15
15. Bill with thin plates on each side.	Lamellirostres.
Bill without thin plates on each side.	Gaviæ.
16. Bill short and hooked.	Psittaci.
Bill rather long, slightly curved.	Cuculi.

ORDER PASSERES.

Feet with three toes in front and one behind; adapted for perching. Bill never raptorial. Primary wing-feathers ten or nine, the secondaries more than six; the wing-coverts short, not more than half the length of the secondaries. Tail usually with twelve feathers.

Sub-order Oscines.

With a complicated vocal apparatus.

Family Corvidae.

Large birds with a stout bill, which is generally notched near the tip. Nostrils usually covered with feathers. Found all over the world except Polynesia.

Genus Glaucopis.

Bill strong, the upper outline much arched. Nostrils partly concealed by feathers. Mouth with wattles, but no bristles. Wing short and rounded, the tip formed by the sixth and seventh quills. The first primary about two-thirds of the second, which is shorter than the secondaries. Tail long and rounded, each feather ending in a blunt point. Legs and feet strong, the tarsi scutellate, longer than the middle toe. New Zealand only. The New Zealand crow is an aberrant member of the Corvidæ, having affinity with the Australian magpies and the bower birds.

Key to the Species.

The wattles entirely blue.	G. wilsoni.
Base of the wattle blue, the rest orange.	G. cinerea.

The North Island Crow.—KOKAKO.

Glaucopis wilsoni.

Similar in colour to the South Island species, but the tail is olivaceous black. Wattle entirely blue. Eye dark brown. Length of the wing, 6 in.; of the tarsus. 2.7 in. Egg—Pale stone grey, spotted with purplish grey; length, 1.45 in. North Island.

Our two crows are numbered among the birds that are retreating before the advance of civilisation. At one time they

E

frequented almost all parts of the country, from the bush-clad mountain ranges to the valleys sloping down to the sea. When Europeans first came here, these birds abounded on Banks Peninsula, but they have now retired to the higher and more remote bushes of the interior. Even there they are seldom seen.

Though at times somewhat shy, they are usually gentle and confident in their manners towards human beings. These characteristics, which to us form one of their greatest charms, are leading them to their destruction. Being easily caught, they rapidly fell a prey to man and to the rats that came in his ships.

They fly about in pairs, and it is thought that they mate for life. The male bird, who is a very pugnacious little fellow, fights with great determination whenever he meets another crow of his sex. But to his mate he is as gentle as a sucking dove. The pairs seem to be mutually and demonstratively affectionate. Male and female have been seen sitting on some fruit-bearing tree, constantly caressing each other with their bills; a pair kept in confinement behaved in this manner for about two years; and, when one of them died, the other survived only a few days.

The note of the New Zealand crow is in pleasing contrast with the harsh and dissonant croak of the Old Country crow. It is remarkably sweet and plaintive, and has been compared to the notes of a flute exquisitely played upon. During the birds' breeding season, their song is one of the most varied and beautiful of all New Zealand bird songs.

In disposition our crows display some of the true characteristics of all the crow family, being inquisitive and crafty; but they differ again from the Old World crows in being poor and feeble flyers, as their flight is limited to 200 yards. They lack the audacity, malice, covetousness, and cupidity of the British and American members of the family, and, so far as is known, they make a proper distinction between "meum" and "tuum." They recognise their human friends, and become much attached to people who take notice of them. There are two species, one for the North Island, the other for the South; but the habits and characteristics of the two species are very much alike, and the Maoris class both under the one name, kokako.

The North Island crow was once common on all the ranges of
the forests; but it now frequents only the higher ranges, away
from human habitations, where it appears at the end of April
and stays till September. It is confined to the North Island.
The male and the female have the same call, consisting of two
notes, like "vio"; but the song of the male consists of five

(*Buller.*)

New Zealand Crows.

pleasant notes, like "vio, ku, ku, ku," which seem to be near,
even when the bird is a considerable distance away.
Though these birds are very tame, they hide in the
thick trees when approached, and at intervals peep
through the branches to see if the intruder has disappeared. When satisfied in this respect, they begin to
whistle. If disturbed a second time, they will seek cover with
marvellous swiftness, hopping through the branches from one
tree to another. If their call is imitated, they will sometimes
come hopping along, so near that a person may almost touch
them. In October, they retreat to their thickly-wooded gullies
between the higher ranges, where they breed, and where they
are not often seen. It is thought that they breed twice a year, and
have two or three young each time. A description of a nest states

that it was built carelessly, of dried branches, ferns, and moss, and was about 30 feet from the ground. The bird feeds mostly on berries, and on the young leaves of various plants, often holding the food in its claw like a parrot. The wings, which are short and small, are seldom used; but the crow is very active and quick in climbing. In the pairing season, when not disturbed, the male makes various evolutions by drooping and spreading his wings, and erecting his tail, while, with bent down head and outstretched neck, he whistles to the female, who remains still and admires his movements.

South Island Crow.—KOKAKO.

Glaucopis cinerea.

Dark bluish grey, with a black band from the nostrils to the eye. Tail blackish at the tip. Wattle orange with a blue base. Length of the wing, 6.25 in.; of the tarsus, 2.7 in. The sexes are alike, but the young are browner in colour, and have smaller wattles. Eye dark brown. Egg—Dark purplish grey, spotted with brown, principally at the larger end; length, 1.6 in. South Island.

The South Island crow, besides frequenting the bush, haunts open places and light scrub. It has been found on the sea shore, and three thousand feet above sea level. It used to be very plentiful in the mountain ranges, and shepherds state that, in severe weather, it comes down to the lowlands. Sometimes the birds may be seen roaming about in pairs with their brood, generally three in number. They are very tame, but, like the North Islanders, they are adepts at the art of hiding when disturbed. It is only when they are in extremities that they attempt to fly. The male and female are inseparable. The male sounds a very sweet whistle, consisting of six notes, such as ''te, to, ta, tu, tu, tu,'' and the call of the female is composed of five, such as ''te, a, tu, tu, tu.''

Mr. Reischek relates that, at Milford Sound in 1884, he shot a crow, and hid himself until its mate appeared. Then he showed himself, and, to his astonishment, the second bird, instead of flying away when it saw the human intruder, went to its companion, hopping round and calling, evidently in a great state of agitation. He sympathised so much with the bird in its bereavement that he let it go. In the south the pairing season begins in October, when the male makes extraordinary evolutions before the female. He bows his head, spreads his wings, and erects and spreads his tail, making at the same time a gurgling noise. In the early days of colonisation, men traversed rough woods in order to find the nests of the South Island crows. They are built in the scrub, not far from the ground, and are composed of twigs and moss. In the beginning of December the

Nest of South Island Crow.

female lays from one to three eggs. The young birds, which are fully grown in May, remain with their parents until the pairing season: they thrive well in confinement, feeding freely on bread and milk, and green stuff, with a few grubs. The nests are easily reached by rats and cats, and, in some localities, where these animals are numerous, the parent birds rarely succeed in rearing a brood. They feed mostly on fruit, berries, and young leaves. The South Island crow is described as a specially beautiful bird in its native haunts. It is so tame that its habits may be studied closely.

Family Turnagridae.

Bill stout, the upper mandible notched near the tip. Wings short, the first primary more than half the length of the second. Nostrils partly covered with feathers. Tarsi scutellate. Mouth-bristles distinct. New Zealand only. This family, which was made by Sir Walter Buller, is related to the thrushes, but the bill is thicker, the wings shorter, and the plumage more lax. Professor Garrod says that it is a non-migratory thrush, which has been modified for a graminivorous diet.

Genus Turnagra.

Bill higher than broad, arched above. Tip of the wing formed by the fourth to the sixth quills, the first about two-thirds the length of the second.

Key to the Species.

Throat olive streaked with white.	T. crassirostris.
Throat white.	T. tanagra.

The South Island Thrush.—Piopio.

Turnagra crassirostris.

Above olive brown; below olivaceous, streaked with white, tinged with yellow on the abdomen. Tail and some of the wing-coverts rufous. Eye yellow. Length of the wing, 5 in.; of the tarsus, 1.25 in. The sexes are alike. The young are rufous on the head and throat. Egg—White, spotted with dark brown; length 1.45 in. South Island.

The North Island Thrush.—-Piopio.

Turnagra tanagra.

Above olive brown. Throat white, breast olivaceous grey, abdomen yellowish white. Tail rufous. Length of the wing 5·25 in. Eye yellow. The sexes are alike. The young birds are marked with rufous on the head, and have a band of the same colour on the breast. North Island.

The native thrushes were among the first birds to beat a hasty retreat to the interior when human beings came to transform the flora; but up in the mountains and in the river gorges, these birds may still be seen hopping about the bush glades, after the manner of thrushes in England.

As early as 1871, Mr. Potts, in some of his charming notes on natural subjects, deplored the fact that the thrush was disappearing from Banks Peninsula and other parts of Canterbury. He saw in this bird a philosopher, with quite as good a claim to the title as that of many human beings to whom it has been accorded. ''Whoever doubts this,'' said the enthusiastic naturalist, ''may make the bird's acquaintance, not merely a slight acquaintance, but one which will ripen into intimacy. If he does this, he will know a bird who takes the world as it is. It is indifferent as to

(*Buller.*)

North Island Thrush.   South Island Thrush.

the nature of its food, feeding on insects when they are procurable, or making shift with grasses, seeds, or fruits. It neither courts nor avoids observation; is as bold as the robin and the tit, but does not possess their intrusive friendliness; when in the presence of strangers, it coolly pursues its occupation, without the prying inquisitiveness of the brown creeper, or the watchful distress of the popokotea; and it defends its home with almost the courage of the falcon or tern.

"It seems to delight in openings found in the river beds between long belts of tutu and other scrub. There it may be seen hopping along the ground, or fluttering about the lower sprays of shrubs, flying out to the spits of sand or drifted trees that lie stranded in the river bed On some of the longer formed spits, that are becoming clothed with vegetation, it searches among the burry acæra, and snips off the fruit stalks of moss, picking the seed of trailing veronicas. It generally goes along the ground in a deliberate manner, and hops with both feet together. With each movement there is a slight flutter of the wings, and a flit of the tail. When it is approached too closely, it leaves its perch, always descending at first, as though safer when near to, or actually on, the ground. Rising on the wing, it gains momentum by a succession of hops. Some of its habits are like those of the wattle-bird, but it generally associates, at least through the summer months, with tuis, parrakeets, and robins. It does not display much secretiveness in choosing a site for its nest, which is usually found seven or eight feet from the ground, but some- times at four feet, and sometimes at over twelve. The nest is firmly and compactly built, with small sprays for the foundation, on which moss is abundantly interwoven with pliant twigs. The lining is generally composed of fine grass bents. Some of the nests are finished off with soft tree-fern down. They are generally placed in the tutu, and sometimes in the coprosma and manuka. Rivals of its own species, as well as other birds, are driven off from the neighbourhood of the nest."

The eye of the thrush seems to gleam with intelligence. This impression is made by its narrow iris of a bright pale yellow. The tongue is pointed, and furnished on one side with a strong muscular apparatus of almost horn-like consistence. Both skin and flesh of the bird are dark; but it is stated that the flavour of the flesh is not at all unpleasant. "It makes a savoury broil for those who bring the proper sauce—hunger," adds Mr. Potts, and he says significantly: "When not so provided, they do wanton mischief who kill birds so harmless and interesting." They are very sociable, and a bush-hand, living the life of a hermit in his whare of tree-fern stems up the Waio River, fed some thrushes until he had enticed them to enter his hut.

An interesting theory has been put forth in regard to the thrush's imitation of the redbill's note. It is regarded as a good instance of the protective mimicry of sound. The thrush gets ample food, in the summer days at least, from the glades of the river-beds. Soaring high above these, the hawk notes the movements of the thrush below, but hearing the simulated cry of the red-bill, withholds his dashing swoop, knowing that a red-bill would alarm his faithful mate, and that the pair, with forces combined, cannot be attacked with impunity

Mr. W W Smith, in an article published in 1888, says that the South Island thrushes were then still fairly numerous at Lake Brunner, but had disappeared from the lower gullies of the Arnold, between Stillwater, now named Richardson, and the Arnold gold diggings. Many years ago the species existed in large numbers on the Maori Gully goldfield, and fed among the huts and tents of the diggers, frequently entering and hopping on the floors, and picking crumbs; but gradually their numbers diminished, until not a single thrush was to be found on the field. Mr. Smith finds that the early morning or evening is the best time to hear this bird's splendid notes and call, or to study its habits, as it is then most active. A few hours after sunset, he says, the thrushes cease to sing or to answer one another's notes, and in fine weather, during the day, they generally remain silent among the fern trees and lower branches of other trees. In dull or wet weather, the birds move about the higher branches in search of food, and avoid the heavy drip of the thick undergrowth.

Like other species, such as the wood-robin, the yellow-breasted tit, the crow, and the weka, the thrush is easily attracted to the spot where any unusual noise is produced in the bush near its haunts, often coming almost within reach of the individual attracting it, spreading its beautiful rich brown tail, moving sideways along the branch, and turning its body right and left, meanwhile examining the stranger closely. It is a powerful flier, and flies with great precision through the tangled vegetation. Mr. Smith has seen it performing these flights: resting almost motionless for some minutes on the high limb of a tree, it would suddenly ruffle its feathers, and, dropping from the limb, fly with great force through the thick undergrowth, reappearing again in the distance on another high limb.   One bird, watched

Nest of South Island Thrush.

by Mr. Smith, uttered a wild jubilant note as it dropped from its perch, to repeat its flight from tree to tree.   He says that this is probably a habit peculiar to the pairing season, as he has never observed them perform such flights at other times of the year.

Captain Cook was much struck with the song of our thrush when he heard it for the first time, and Sir Walter Buller describes the North Island bird as the best of the native songsters. The song, he says, consists of five distinct bars.   Each is repeated six or seven times in succession; but the bird often stops abruptly in its overture to introduce a variety of other notes, one of which is a peculiar rattling sound, accompanied by the spreading of the tail, and, apparently, expressive of ecstacy. Its most common Maori name is piopio, which has been given to it on account of its peculiar whistle. The northern and southern species are confined to their respective islands. Nests belonging to the latter are displayed in the Canterbury Museum. One was obtained close to the Waio River, in Westland, and is loosely made of dry twigs, pieces of bark, moss, and grass. It was constructed in the fork of a tree.

Family Sylviidae.

Small birds with a slender bill, shorter than the head; the mouth bristly, and the nostrils exposed. Wing with ten primaries, the first about half the length of the second. Tarsi booted; that is, there are no divisions on the front surface. Distribution universal, but especially common in the eastern hemisphere.

Key to the Genera.

1. Tail nearly as long as the wing.	Pseudogerygone.
Tail much shorter than the wing.	2
2. Second primary feather longer than the secondaries.	Petrœca.
Second primary about equal to the secondaries.	Miro.

Genus Pseudogerygone.

Tail nearly as long as the wing, rounded. Second primary about equal to the secondaries. Tarsi moderate, longer than the middle toe. Australia and New Guinea to New Zealand.

The Grey Warbler.—Riro-riro and Horirerire.

Pseudogerygone igata.

Above, greyish olive; throat and breast grey, abdomen white, tinged with yellow on the abdomen. Tail black, with a white spot near the tip of the lateral feathers. Eye red. The sexes are alike, but in the young there is no yellow on the lower surface. Length of the wing, 2 in.; of the tarsus, 0.75 in. Egg—White, or pinkish spotted with red, principally at the larger end; occasionally pure white. Both islands of New Zealand.

The grey warbler, which is found in all parts of New Zealand, is the object of much commiseration, and it is the principal victim of the cuckoos' parasitical habits, described further on. In spite

(Voy. Erebus and Terror.)
Chatham Island Warbler.

of the injustice under which it labours, and the usurpation of its nest, it maintains a very cheerful aspect. Fortunately, the warbler is not of a worrying disposition, and it apparently does not resent the cuckoos' intrusion.

It is the most active, lively, and industrious of all our birds. Its sweet notes and merry song are heard all day long, as it flits from branch to branch. In warbling its welcome notes, it often spreads its white-tipped tail so as to form a fan. It is one of the earliest breeders. Its pensile nest, which is a wonderful structure, is dome-shaped or pear-shaped, and may often be found in August hanging in bushy manuka or olearia. The nest is suspended by its top, and is kept from swaying in the breeze by slight fastenings to a spray or two, which act as guys. Moss, and occasionally wool, are largely used for building material, which is sometimes bound into a compact form by cobweb. Poultry-feathers, down, pieces of string, and cotton are also used when available. There is a small entrance near the middle, surmounted with a kind of porch. The interior is thickly lined with feathers. In some instances the nest is built without either dome or porch, and, if the site chosen calls for a change

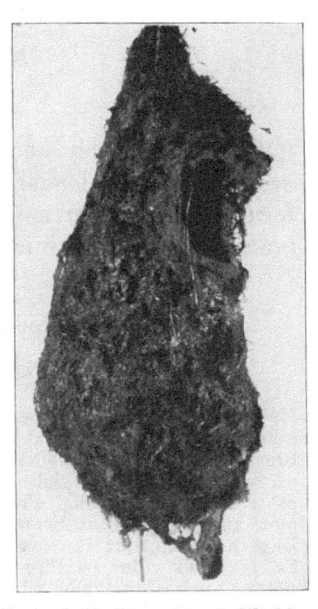

Nest of Chatham Island Warbler.

in the principle of construction, Mr. Potts says, the nest is not suspended. There are generally five or six eggs in the nest. The breeding season commences early, and extends through the summer.

The Chatham Island Warbler.

Pseudogerygone albofrontata.

This bird differs from the previous one in having the forehead and throat white. Eye light red. Length of the wing, 2.6 in.; of the tarsus, 0.9 in.; Chatham Islands. Egg—Pinkish white, with numerous red spots and lines. Length, 0.74 in.

The Chatham Island warbler is slightly larger than the New Zealand species. Its nest is more closely felted than that of the grey warbler, and is composed mostly of small fern-roots, moss, dead leaves, lime, soft grass, and spider's webs. The nest is also more rounded in shape, and is stronger. There are generally four eggs.

New Zealand Tits.

Genus Petroeca.

Tail considerably shorter than the wing, square at the end. Second primary longer than the secondaries; tip of the wing formed by the fourth to the sixth quills. Feet moderate. The tarsus less than the middle toe and claw.

The Yellow-breasted Tit.—NGIRU-NGIRU.

Petroeca macrocephala.

Male—Head, neck, and back black, with a white spot over the bill: below yellow, bright on the breast, getting paler on the abdomen. Wing, brownish black; some of the primaries and the secondaries with a band of white. Tail, blackish brown, the three outer feathers with a band of white. Eye black. Female—Above brown, with a small white spot over the bill; throat brownish white, abdomen yellow. Wings and tail like the male, but the white on the wings tinged with yellow. Young— In each sex coloured like the adult, but the tints are not so pure, and paler. Length of the wing, 2.9 in.; that of the tarsus, 0.9 in. The female is rather smaller. Egg—White, with spots of purplish grey, often forming a ring near the larger end. Length, 0.75 in. South Island of New Zealand, Chatham Islands, Auckland Islands.

The New Zealand tits, like those of the Old Country, are bold to the extent of impudence, but they are also very friendly and sociable. They are perpetual motion personified, as they are never at rest.

The southern, or yellow-breasted, tits are sprightly and graceful little birds. "They may often be seen playing together," says Mr. W. W. Smith, "and chasing each other through the branches, gently fluttering their wings, erecting their crests, and uttering a suppressed twitter, as they sit eyeing each other on the boughs, or clinging to the stems of the trees, and exhibiting the peculiar jealousy of the wood robin about the tents.''

Yellow-breasted Tit.   White-breasted Tit.

(Buller.)

North Island Wood Robin.   South Island Wood Robin.

The female undertakes most of the work of building the nest, which is composed mostly of mosses, grass-bents, very thin and fine sprays, cobwebs, woolly scales of fern-trees, and dead leaves, while the interior is lined with feathers. There are generally four eggs. Nests have been found in porches, sheds, and outhouses, on the face of a high cliff by the sea, under the arch of a bridge on a public road, in the crevices of cave-like rocks, in the hollows of decayed trees, amongst the pendant leaves of cabbage-trees, in manuka, and in other places. The tit, in fact, seems to have no fixed idea in regard to a suitable site for its

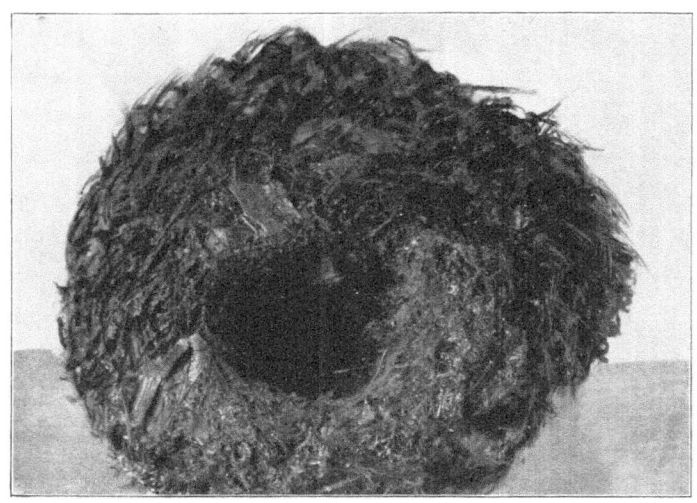

Nest of the Yellow-breasted Tit.

home. It frequents gardens, and is often seen perched on a bough watching for grubs. As an insect-eater, it is very useful.

These birds are extremely quarrelsome. Two combatants have been separated by hand before the victor in an encounter could be induced to leave his panting adversary On one occasion, two were taken by hand whilst fighting on a lawn, and, like a pair of naughty boys, were carried into the house; on being released out of doors, hostilities were immediately begun again.

The White-breasted Tit.—MIRO-MIRO.

Petroeca toi-toi.

Like the last, but with less white on the quills, and pure white on the breast and abdomen. Eye black. Length of the wing, 2.7 in.;

of the tarsus, 0.75 in. Egg—White, speckled with brownish grey.
Length, 0.78 in. North Island of New Zealand.

The North Island species may be seen in the ranges, and on
some of the islands. These beautiful birds favour the clearings.
They hop about on low branches, with wings slightly drooping,
and tails erect, while they utter a sweet whistle of one note, like
"chee," the male, however, sometimes giving the five notes.
The female, being of a retiring disposition, is not seen so often
as the male.

Nest of White-breasted Tit.

They feed on insects and larvæ, and do much good in
destroying large numbers of the former. They have been known
to carry to their nests insects over an inch long.

Both birds build their nest together, using moss, spiderwebs,
and rotten wood, and lining the inside with the down of seeds
taken from flowers. In October the female lays three or four
eggs, and both birds breed and rear the young twice a year.

As the young are full grown in the beginning of December,
the parents then leave them. The nest is flattened in the fork
of a tree, and the site selected is generally in very thick scrub.

New Zealand Wood Robins.

Genus Miro.

Tail considerably shorter than the wing, square at the end.
Second primary about equal to the secondaries; tip of the wing

F

formed by the fourth to the sixth primaries. Feet large, the tarsus longer than the outer toe with claw.

Key to the Species.

1. Abdomen and frontal spot white.		2
Entirely black.		3
2. Legs and feet dark brown.		M. albifrons.
Legs and feet light brown.		M. australis.
3. Legs and feet brown.		M. traversi.
Legs and feet black.		M. dannefordi.

The South Island Wood Robin.—Toutouwai.

Miro albifrons.

Greyish black, with a small white spot over the bill, and a broad band of yellowish white on the breast and abdomen. Legs and feet brown, the soles yellow. Eye black. Length of the wing, 4 in.; of the tarsus, 1.5 in. The female is rather smaller than the male, and greyish brown in colour. The young is coloured like the female, and the white spot over the bill is obscure. Egg—Dull white, with greyish brown marks, principally at the larger end. Length 1 in.

Our wood robins are tame, pert, merry, and cheerful birds, and they are very partial to crumbs. In fact, in all respects except colour they resemble the famous robin redbreast, the English favourite. There is one species for the North Island, and another for the South Island. The list is complete with two jet black species, one of which is found in the Island of Mangare, of the Chatham group, and the other on the Snares.

The South Island robins may often be seen in the bush. They enter tents, hop about inside, and sometimes partake of meals on the table. They are active little fellows. They appear at the break of day, and retire when darkness comes on. They have been known to sit on a man's body, as he lay on the ground, hopping about and picking at his watch-chain or at any other object that attracted their attention. When supplied with a greater quantity of food than they can eat at a time, they sometimes hide a store, and, no doubt, return to it later on.

The nest is described as a complete study in natural history. Boldness of conception is disclosed not only by the scale on which the lines of the ample dwelling are founded, but also by the design, which is very well carried out for the comfortable accommodation of at most four nestlings. The nest is often placed amongst the roots of a large tree near a creek, on a mossy protuberance on a rugged stem, or it may neatly fill a hollow, matching so well the moss-tinted russet brown bark, with its cleverly selected materials, that it is difficult for the eye to detect the bird's home.

Nest of South Island Wood Robin.

The North Island Wood Robin.—TOUTOUWAI.

Miro australis.

Greyish black, with a small white spot over the bill, and a band of white on the breast and abdomen; the shafts of the feathers greyish white; legs and feet pale brown, the soles yellow. Eye black. Length of the wing 3.7 in.; of the tarsus, 1.4 in. The female and young resemble those of the last species, but they can be distinguished by the lighter colour of the legs and feet. Egg—Dull white, with purplish brown spots, which generally form a circle round the thick end. Length, 0.95 in.

The North Island robin is one of the birds that have decreased largely in the north. I had a brief acquaintance with this bird on the Little Barrier Island in February, 1907. The first meeting took place in the bed of a sequestered creek, amongst boulders and tree-ferns and the decaying trunks of mighty forest trees, which had crashed down from the hillside, and had fallen prone across the banks. I had sat on a boulder under the fronds of a tree-fern to rest. Heavy rain had fallen in the night. As the morning was still young, all the trees and shrubs were dripping

wet, and beads hung from almost every leaf. Just as I was rising
to start away again, a shower of water fell upon me. Glancing
up, I saw one of the fronds of the tree-fern moving gently in the
air. A little slaty-grey bird was sitting near the end of the
frond swinging to and fro upon it and watching me with a
pair of jet black eyes, which sparkled like diamonds. His head
was cocked on one side, but he looked me straight in the face,
with an inquisitive and rather quizzical but tolerantly good-
natured expression, as if he knew that I was a stranger and
wondered what I was doing on his property, but wished to set
me at my ease. He displayed absolutely no fear. He showed,
indeed, that he desired to make my acquaintance. Hopping
nimbly off the frond, he sent a shower of water down again.
He hopped on to a twig within a few feet of me. A few more
hops, first to one side and then to the other, brought him closer,
and from that standpoint he looked at me for several seconds.
Having satisfied his curiosity, he hopped away from bough to
bough, until he hopped himself on to the trunk of a tree, where
he hopped up and down, pecking at the bark and moss in search
of insect food. He made five or six little hops, then stopped to
peck at the tree, and listened intently for some sound. He
accompanied me for several yards as I went down the creek,
hopping cheerfully along and peeping out at me from the foliage.
The rapidity with which this, the gentlest of birds and the
kindest and brightest of companions, has decreased from many
districts in the north shows that none needs sanctuary more than
he does. He is not at all rare on the sanctuary. He may also be
seen in large numbers on Kapiti Island, close to the west coast
of Wellington, and in districts where the forest trees are
still standing.*

The Black Wood Robin.

Miro traversi.

Jet black, the wings brownish. Legs and feet brown, the soles yellow.
Length of the wing, 3.3 in.; of the tarsus, 1.1 in. Mangare, one of the
Chatham Group.

---

*By J. Drummond, in the Sydney *Morning Herald*, June 1st, 1907.

The Little Wood Robin.

Miro dannefordi.

Jet black, the wings brownish. Legs and feet black, the soles bright orange, as is also the rictal membrane of the mouth. Eye black. Length of the wing, 2.75 in.; of the tarsus, 0.9 in. The Snares (south of New Zealand).

Family Muscicapidae.

The fly-catchers form a group of small birds with a well-marked notch on the bill, numerous bristles round the mouth, and weak legs and feet. They are found only in the eastern hemisphere, being replaced in America by the tyrant-flycatchers.

Genus Rhipidura.

Wings very short, the first primary about one-half the length of the second, which is equal to the secondaries. Tail very long, and rounded at the end, the two centre feathers only slightly longer than the next. Bill broad at the base, shorter than the mouth bristles. Tarsi scutellate. From India through the Malayan Archipelago to Australia and Polynesia.

The Pied Fantail.—TIWAKAWAKA.

Rhipidura flabellifera.

Head and neck blackish grey, with white throat and eyebrows. The back brown, the breast and abdomen yellowish rufous. The two middle tail feathers black with white tips, the outer ones white, the intermediate white with the outer webs partly black, all the shafts white. Eye black. Length of the wing, 3 in.; of the tarsus, 0.7 in. The sexes are similar. In the young, the upper surface is more or less shaded with rufous, and the lower surface with tawny. Egg—White with brownish grey spots towards the larger end; length, 0.7 in. Both Islands and the Chatham Islands.

Everybody admits that the fantails are among the prettiest and most engaging birds New Zealand possesses. There are two species, and their habits are very much alike, so that what is said of the pied fantail may be applied to the black fantail.

In the first place, they are very industrious birds. This is clearly shown when they are building their nests, which are found near large supplies of insects. A bough standing out from a high bank of a shady creek is a favourite spot. The foundation of the nest is laid by adroitly securing slender chips of decayed wood with lines of cobweb to the spray selected. This delicate operation must necessarily be a work of great difficulty. In places where splinters of decayed wood are not obtainable, pieces of coarse grass have been used instead. Fine grasses, thread -like roots, dead leaves or skeletons of leaves, hair, green tufts of moss, and the down of ferns are also brought into requisition.

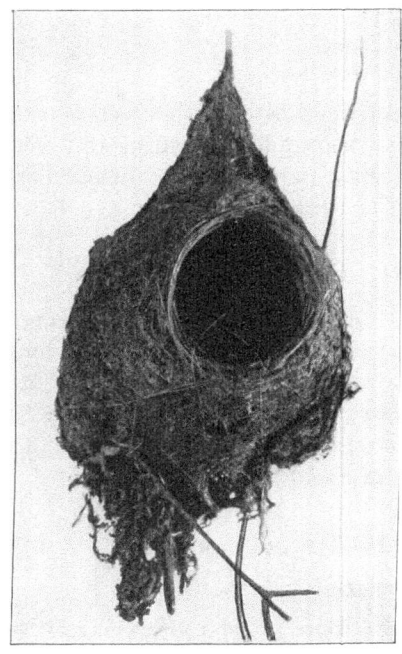

Nest of Pied Fantail.

Both male and female take part in the work, but the latter seems to bear the larger share. The strength of the structure is tested in many ways. It is trampled down, and the webs are carried from the interior to the outside in festoons from left to right, and right to left, as far down as the beak can reach.

The wall gradually rises, and the bird, with elevated tail, is itself the mould by which the rounded cavity is shaped. "Seated in the centre of the rising structure, it turns repeatedly, fluttering its wings, and keeping the wall pressed out to its proper shape. Its head and chin are pressed on to the top, and the materials are pulled in towards the centre. This manœuvre is repeated at frequent intervals. So earnest are the workers, that for hours together they scarcely rest. Sometimes, by a sudden flutter, they obtain a few insects, or the creek is visited for water. The male now and then finds time for a brief twitter, moving his head from side to side, as if criticising or admiring the results of their united efforts, but both are quickly at work again.''

Fantails have been known to frequent houses, clearing the roofs of flies, catching them on the wing, or picking them off the curtains, ceilings, and other places.

The young males, like those of many other species, are low-voiced at first, but improve by perseverance. Mr. Potts, from whom the above description has been taken, states that the fantail is one of the latest birds to retire to roost, it being almost dark when the last feeble twitter sounds through the trees.

A very interesting feature in connection with these birds is the construction of union nests. It is well known that the pied and the black fantails pair together, and that the union is fruitful. But the young, it is stated, never resemble partly one parent and partly the other, as is commonly the case with hybrids. Sometimes the young in a union nest are all pied, and sometimes there are several black ones. The pied are more numerous than the black, but all, apparently, are pure-bred.

The Black Fantail.—Tiwakawaka.

Rhipidura fuliginosa.

Head and neck greyish black, with a white spot over each ear. Rest of the body dark chocolate brown. Quills dark brown. Tail black. Eye black. Length of the wing, 3 in. of the tarsus, 0·7 in. In the young birds the white spot over the ear is absent. Egg—The same as in the last species. Found in both islands, but very rare in the North.

Family Sturnidae.

Medium-sized birds, with an acute conical bill, which is longer than the head, and not notched at the tip; when open, the line of commissure is seen to be angulated. There are no mouth bristles.

Genus Heteralocha.

Bill nearly straight in the male, very long and curved in the female. Nostrils partly covered with feathers. Tip of the wing formed by fifth to seventh quills. Tarsi scutellated, much longer than the middle toe. New Zealand only.

(Nature.)

Huia: male and female.

The Huia.

Heteralocha acutirostris.

Greenish black, with a band of white at the end of the tail. Bill white, wattles orange. Eye dark brown. Length of the wing, 8 in.; of the tarsus, 3 in. Egg—Greyish white, with a few spots of purple and brown; length 1.8 in. Found only in the mountains between Wellington and Hawke's Bay.

The huias are among the remarkable birds that make the fauna of New Zealand specially interesting. Their most striking peculiarity to a naturalist lies in the difference between the bills of the male and the female. The male's bill is straight, while the female's is strikingly curved, and has a considerable advantage in regard to length.

The difference is so marked that, when these birds were first made known to science, many ornithologists could hardly believe that they belonged to one species. Mr. J. Gould, the author of *Birds of Australia,* concluded from the formation of the bills that there were two distinct species, and he gave them distinct names, but he afterwards rectified his error. Another writer, Dr. Sclater, states that such a divergence in the structure of the beak of the two sexes is very uncommon, and is scarcely paralleled in the class of birds. He adds that it is difficult to guess at the reason for the anomaly, or to explain it on Darwinian or any other principles.

Another interesting feature about the huias is the fact, well known to almost all residents of New Zealand, that the Maoris once used the ivory-tipped tail feathers as head-ornaments, denoting tribal rank, such as that of rangatira. Several specimens of the species were obtained by Dr. Dieffenbach when he was in the colony. Writing in 1836, he said that the tail feathers were then in great request among the Maoris in the districts frequented by the bird. The coveted emblems of rank were often sent as presents to friends in all parts of the North Island. "These fine birds," he wrote, "can only be obtained with the help of a native, who calls them with a shrill and long-continued whistle, resembling the sound of the native game of the species. After an extensive journey in the hilly forest in search of them, I had at last the pleasure of seeing four alight on the lower branches of the trees near which the native accompanying me stood. They came as quick as lightning, descending from branch to branch, spreading out their tails, and throwing out their wings. Anxious to obtain them, I fired; but they generally come so near that the natives kill them with sticks."

In olden times, some of the Maoris seem to have given a considerable portion of their time to capturing huias. The Rev. W. Colenso, in one of his articles on the Maoris, refers to an eccentric old chief and tohunga, named Pipimoho, who lived near Castle Point. He was the only person in those parts who knew how and where to capture huias. For a long time he had

no other occupation. Regularly once or twice a year he went to
the inland forests from the East Coast to obtain the precious
birds, and supply tail feathers to the principal chiefs in Hawke's
Bay, his superiors in rank. The Maoris have been known to keep
the huias in captivity, placing them in a large light cage of
network. They were fed with soft vegetable food, and, when
they arrived at maturity, their tail feathers were plucked. A
complete apparatus, consisting of a long rod, a string, and a

Nest of the Huia.

running knot, was used for bird-snaring operations. The female
huia was generally taken first, the male being more easily snared
after its mate had been captured.

These birds are quiet, amiable, sociable, affectionate to each
other, and very tame; and they seem to thrive well in captivity.
As the male and female are generally found together, it is
concluded that they are monogamous. They can adapt themselves to animal or vegetable food, and, in obtaining supplies,
they are mutually helpful.

Mr. Colenso once received from a settler the head of a female
huia with a remarkably formed bill. It looked like a large cork-
screw, having a spiral of two large and equal curves. It is
thought that the bird was hatched with this peculiarity, which
made the bill worse than useless, as it was an obstacle to eating,
and kept the mouth always open. The bird was an old one. The
fact that it lived in spite of its deformity is explained by the

supposition that the unfortunate possessor of the bill was helped throughout a long life by a kind and attentive mate.

Huias live together in the forests, subsist partly on insects, and partly on small fruits, and are seldom found far from the Ruahine, Tararua, and Rimutaka Ranges, in the North Island. Being very bad fliers, they rarely attempt to fly. Their black plumage, with its greenish gloss, is in striking contrast to the orange wattles, ivory bills, and white-tipped tail feathers. They have been referred to the hoopoes, solitary, songless birds that live for preference in low and humid places, and pass nearly all their existence on the ground, and also to the crows. It is now generally recognised, however, that the huias are most intimately related to the starlings.

They build their nests in hollow trees, and line them with grass and twigs and similar substances. An observer who resided near Woodville has given an interesting account of a family of four young huias, which were so nearly of the adult size as to be scarcely distinguishable from the parent birds, which, when observed, were busily engaged in feeding them. Their activity was remarkable, and the speed with which they traversed the wood, hopping, or rather, bounding, with a slight opening motion of the wings, flying very short distances, was speedily noticeable. Owing to the moist nature of the locality, the huge trees were clothed in mosses and parasitic ferns. The huias constantly ripped into strips these mosses and also small climbing ferns, fragments of which were seen dropping down from the branches where these beautiful birds were zealously working for their young.

Genus Creadion.

Bill nearly straight, longer than the head; the nostrils covered with feathers; the mouth carunculated. The wings short and rounded, the tip formed by the fourth to the sixth feathers. Tarsi scutellated, not quite as long as the middle toe. New Zealand only.

The Saddle-back.—Tieke.

Creadion carunculatus.

Black, with the back and wing coverts deep chestnut. Wattle generally orange, but variable between red and yellow. Eye dark

Saddle-back: old and young.   *(Buller.)*

brown. The sexes are alike, but the young is brown, with the lower part of the back dull chestnut. Length of the wing, 3.5 in.; of the tarsus, 1.6 in. Both Islands. Egg—White, spotted with brownish grey and violet; length, 1.15 in.

The saddle-back is a noisy, chattering, and amiable bird, which seems to regard its little fellow creatures with great good will. A shrill note, unlike that of any other bird, repeated several times in quick succession, announces its sudden but welcome appearance. Its movements are notably prompt, rapid, and decided, and "no sooner has it sounded its call-note," Mr. Potts says, "than it emerges from its leafy screen, and bounds before the spectator as suddenly as a harlequin in a pantomime. From these abrupt movements, or flying leaps, it seems to perform a role of its own, that appears almost startling amidst the umbrageous serenity of the forest. Let the eye follow its motions, which are so quickly changed, and watch the saddle-back perched for a few moments on the lichen-mottled bole of some fallen tree, its favourite position. The glossy black plumage is relieved from sameness by the quaint saddlemark that crosses the back and wings. The observer will probably notice that its attitude is peculiar; to use a colonial phrase, 'it has a strange set about it.' The head and the tail are rather elevated; the feathers of the tail take a gently sweeping curve; and the bird looks as though it is prepared to leap. One more glance, and it is away, climbing some moss-clothed trunk, or picking its food from beneath the flakes and ragged strips of bark that hang from the brown-stemmed fuchsia tree."

The saddle-backs have a curious habit of sometimes following flights of canaries, or yellowheads, through the bush. Apparently they act as scouts to the smaller birds. A very pretty theory has been built up in connection with this habit. But it may fairly be assumed that the real attraction of the saddle-backs is not of a sentimental nature, but is merely a desire to obtain supplies of food. Apart from the motives, good or bad, that actuate the saddle-backs, they present a charming sight when attending the flocks of little yellowheads in their journeys through the forests.

Mr. W. W. Smith has given a delightful glimpse into a woodland scene near Lake Brunner, in which saddle-backs and yellowheads were the principal actors. He says: "I had travelled on the banks and bed of a creek for almost a mile, when I turned to the right up a small narrow gully in search of ferns

and other botanical rareties. On reaching nearly the top of the
gully, I heard the shrill ringing notes of a flock of yellowheads.
As I noticed them crossing the gully some distance above me,
I moved on gently, until I was under the branches to which the
birds were advancing from the gully. They numbered about
200, and were in rich plumage. They fed eagerly for some
minutes among the branches of the trees, and then, simultaneously uttering their call, they flew forward some yards and
began to feed, until they again sounded the signal to advance,
repeating it at short intervals, and passing on through the bush
in this order.

Nest of the Saddle-back.

"Before the yellowheads had quite disappeared, I heard the
rich flute-notes of a flock of saddle-backs advancing. I climbed
up the side of the gully, and stood on the edge. Two males
were the first to appear, and they were followed by the remainder
of the flock. They advanced in the line of the yellowheads.
When they noticed me, some approached closely, twittering and
elevating their tails. They moved about in a sprightly manner
on the lower branches, within a few feet of my face, scanning
me carefully, and wondering, perhaps, at the intruder into their
solitary domain. They were exceedingly tame, and moved with
great activity. stopping at intervals, and resting their breasts
for a few seconds on the boughs, and again proceeding, searching
carefully for food among the ferns and mosses. They remained

for about seven minutes, and then disappeared slowly in the track of the yellowheads."

In the breeding season, when the female is hatching, the male often sits on a branch near her and sings, and his notes are not so harsh then as on other occasions. If the saddle-back notices anything unusual, it hops about in a very excited manner, with the wings close to the body, and the head bent downward, and it stops and listens at intervals. When it has satisfied its curiosity, it flutters to a distance. As its wings are weak, it cannot fly far. It is sometimes seen in the open parts of the forest, but prefers dark, damp, shady places.

Family Timeliidae.

Bill slender, not notched at the end. Wings short, rounded, concave; with ten primaries, the first much shorter than the second, which is about as long as the secondaries. The babbling thrush (*Timelia*) of India, forms the type of this family, but it is made to include other Old-world forms. It is not found in America.

Genus Sphenaeacus.

Bill short, higher than broad at the nostrils; no mouth bristles. Tip of the wing formed by the fourth and fifth feathers. Tail feathers ten, long, graduated, with spiny shafts and loose webs. South Africa and New Zealand.

Key to the Species.

1. Throat white.	S. rufescens
Throat spotted.	2
2. Darker and smaller.	S. punctatus
Lighter and larger.	S. fulvus.

The Fern-bird.—MATATA and TOETOE.

Sphenaeacus punctatus.

Above, fulvous-brown; forehead rufous with a black streak in the middle of each feather; a pale streak from the nostrils over each eye. Below white, spotted with black; tinged with fulvous on the flanks and abdomen. Eye black. Length of the wing 2.3 in.; of the tarsus, 0.8 in Egg—White, speckled all over with violet and greyish red; length 0.8 in. Both islands.

Fern-Bird.

Chatham Island Fern-Bird.

(Buller)

The fern-bird, as its name implies, frequents ferny lands. Mr. H. Guthrie-Smith of Tutira, Hawke's Bay, who has seen this bird on his estate, says that it is generally a shy bird; but, in spring, the male seems to lose a great deal of his timidity, and, regardless of the presence of man, will mount to the top of a flax stick, climbing up with little runs, like a mouse or house-fly, his tail being bent in towards the stem.

This species has been called the grass-bird, and also the utick, on account of its peculiar cry. As early as 1870, Mr. Potts recorded the fact that the bird was becoming rare in Canterbury, and was rapidly disappearing as the swamps were drained. Its monotonous note, which at one time was heard in almost any place where the tall toi-toi reeds or the waving leaves of the Carex virgata grow, indicates marshy ground. Being feeble fliers, these birds perished in large numbers in the bush fires when lands were cleared by the settlers. The nest is somewhat oval in shape, and measures about three inches across. One specimen preserved is so frail that it may be seen through. The bird usually builds in a tussock, a few inches above the level of the ground.

The Tawny Fern-bird.

Sphenaeacus fulvus.

Paler than the last, the brown centres to the feathers being narrower. Length of the wing, 2.5 in.; of the tarsus, 0.9 in. South Island and the Snares.

The Chatham Island Fern-bird.

Sphenaeacus rufescens.

Upper surface, as well as the wings and tail, rufous brown; lower surface brownish white, the throat tinged with yellow. A brownish white streak over the eye; the sides of the head marked with black. Length of the wing, 2.5 in.; of the tarsus, 1.0 in.—Egg—Creamy white, speckled and marbled with reddish brown; length, 0.8 in. Mangare, one of the Chatham group.

Mangare, in the Chatham group, is a very small islet, and has a stony surface, which, however, is nearly covered with low rigid scrub. When Mr. H. H. Travers visited the island, in 1872, he found that the fern-bird was not uncommon. Its habit of hopping rapidly from one point of cover to another increased the difficulty of securing the bird. It has a peculiar whistle, Mr. Travers says, very like that which a person

G

uses in order to attract the attention of some other person at a distance; although Mr. Travers knew he was alone on the island, he frequently stopped mechanically on hearing the note of the bird, under the momentary impression that someone was whistling to him. This bird is solitary in its habits, and seems to live exclusively on insects.

Nest of Fern-bird.

Family Paridae.

Bill rather short, not notched at the tip. Nostrils partly concealed by tufts of feathers. Wing with ten primaries, the first much shorter than the second. Tarsi scutellate.

Key to the Genera.

1. Tail shorter than the wings, the shafts of the feathers
   projecting.   2
   Tail longer than the wings, the shafts of the feathers
   not projecting.   Finschia.
2. Hind claw longer than bill.   Mohua.
   Hind claw shorter than the bill.   Certhiparus.

These genera belong to the Paridæ australes, which are confined to the Australian Region. The nostrils are longitudinal, and are not completely covered by feathers.

Genus Mohua.

First primary about two-thirds of the second. Tail feathers frayed at ends, the shafts projecting. Tarsi with three or four scutellæ, which are often united. Feet large, the middle toe and claw nearly as long as the tarsus; hind claw longer than the bill. New Zealand only

The Bush Canary.—MOHUA.

Mohua ochrocephala.

Head, breast, and abdomen, yellow; back, yellowish brown; tail, yellowish olivaceous. Wing-feathers, brown, mostly edged with yellowish brown. Eye black. Length of the wing, 3.2 in.; of the tarsus, 1 in. Egg—White, with small faint specks of red; length, 0.9 in. The sexes are alike; the young are much duller in colour. South Island only.

Nest of Bush Canary.

The New Zealand canary has a sharp, strident call, and its movements are quick and active. Its golden head and breast are seen popping out of the green foliage, and as suddenly disappearing again. The bird may often be seen on the ground, searching amongst the moss for insects. The nest is made principally of moss, tightly pressed together, and interwoven with spider webs. It may be found in the hollow trunk of the broad-leaf, and sometimes in a decaying black beech.

Genus Certhiparus.

First primary about two-thirds of the second. Tail feathers frayed at the ends, the shafts projecting. Tarsi with five or six scutellæ, which are never united. Feet moderate, the middle toe and claw four-fifths of the tarsus; hind claw shorter than the bill. New Zealand only.

The Whitehead.—POPOKOTEA.

Certhiparus albicapillus.

Head, breast, and abdomen brownish white. Back and tail brown. Wing feathers dark brown. Eye black. Length of the wing, 2.8 in.; of the tarsus, 1 in. Egg—White, faintly speckled with pink; length, 0.9 in. The sexes are alike. In the young the head and lower surface are greyish white. North Island only.

Whiteheads belong exclusively to the North Island, but they have close relatives in the South, with the same habits and characteristics, but with bright yellow caps instead of white ones. The two species are popularly classed together as "bush canaries." I had hardly entered the bush on the slopes of a hill near the landing-place at Kapiti Island when I heard a noisy twittering in the distance. It was indistinct at first, and could barely be heard above the rustling of the leaves in the breeze. It came nearer and grew louder, until the higher branches seemed to be filled with it. Then a dozen little birds flitted into view They were the busiest birds imaginable. Their affairs seemed to be of the utmost urgency. The excitement that prevailed was intense. They were never silent and never at rest. They turned completely round with one hop off the branch, and with another hop they turned back again. I watched them for several minutes, trying in vain to fathom the cause of their activity. They had no time, apparently, to feed upon the insects on the tree, but as soon as they saw me underneath, they peeped down over the branches and through the leaves, and gave many signs of the interest and curiosity my presence had aroused in their breasts.

Their cries have very little variety, but are not at all monotonous. They are like, "Chee, chee, chee; chee, chee, chee; chee, chee, chee," repeated time after time, without end. By placing the back of my hand to my lips, and sucking in my lips with a labial sound, I made a fairly good imitation of their cries. This brought them nearer, and some of them came so close that I could almost have stretched out my hand and touched them.

Whitehead.

(*Voy. Erebus and Terror.*)
Brown Creeper.

While I was watching their strange antics, the twitter suddenly stopped for a few seconds. When it began again, it was dying away in the distance over the tree-tops. The whole flock had suddenly thought of some urgent business in another part of the forest, and had set out post haste to attend to it. I saw very many whiteheads on the sanctuary afterwards, but I never saw one passing the time in idleness. I never knew one to be silent for more than a few seconds, and I never saw any signs of fear amongst them at the presence of a human being. The rather oppressive silence of the dark and gloomy forests was broken more frequently by whiteheads than by any other birds. Sometimes one of them would raise its voice out of the usual twitter, and would give expression to a fairly loud and very pleasing

Nest of Whitehead.

trill. This was done generally when two were together, although, as far as I could observe, the trill was given by only one of the pair. The whitehead is one of the native birds whose doom has been announced, but it is satisfactory to learn that it is plentiful on both the Little Barrier and Kapiti sanctuaries, and, according to recent information, in many places on the mainland of the North Island. It affords one of the pleasantest scenes of the forest. When all other birds are absent or silent, it comes along and enlivens the forest with its busy twitter and its extraordinarily quick movements. It is not a pretty bird. Its prevailing colour is white, often a dirty white, tinged with brown, but it has a neat and tidy appearance, and its charming disposition makes it one of the greatest favourites of the forest.*

---

*By J. Drummond, in an article on " Kapiti Island," *Lyttelton Times*, March 21st, 1908

Genus Finschia.

First primary about one half the length of the second, the tip of the wing formed by the fourth to the sixth quills. Tail graduated, the shafts of the feathers not projecting. Legs and feet slender; from three to eight scutellæ on the tarsus. New Zealand only.

Nest of Brown Creeper.

The Brown Creeper.—PIPIRIHIKA.

Finschia novae-zealandiae.

Head, back, and tail reddish brown; sides of the head and nape dark grey; under surface fawn. Lateral feathers of the tail with a broad brown spot. Eye grey. Length of the wing, 2.5 in.; of the tarsus, 0.8 in. Egg—White, with grey and brown spots, which form a ring round the larger end; length, 0.7 in. South Island only. The sexes are alike; and the young are darker below. Some of the birds from the West Coast Sounds have a white eye-brow.

The brown creeper is seen in the bush lands. A nest obtained by Mr. Potts far above the Rangitata Gorge contained three young birds. It was compactly built of moss, with a few feathers, and had been placed in a beech tree, between the trunk and spur, about seven feet from the ground.

Family Motacillidae.

Bill slender, straight. Wing with only nine primaries, the first fully developed; the inner secondaries lengthened. Tarsi scutellate, the toes lengthened.

Genus Anthus.

Wing with the first three primaries equal and longest. Tail emarginate. Tarsi longer than the middle toe ; lateral toes equal ; hind claw very long. Found all over the world except Polynesia.

(Meyer.)
Ground Lark.

The Ground Lark.—PIHOIHOI.

Anthus novae-zealandiae.

Above brown, below dull white, with brown spots on the breast. Cheeks white with brown spots; a dark brown band through the eye. Outer tail feather white except near the base; the shaft white. Second feather white with an oblique mark of brown on the inner web; the shaft brown. Hind claw equal to hind toe. Second, third, and fourth primaries sinuated on the outer margin. Eye dark-brown. Length of the wing 3.6 in.; of the tarsus, 1 in. Egg—Greyish white, speckled all over with greyish brown; length 0.7 in. The sexes are alike. In the young the feathers of the upper surface, especially the wing coverts, are margined with brownish white. Both islands, Chatham Islands, and Auckland Islands.

The New Zealand lark is a well known bird. Its nest, which is on the ground, is made of grass, and is generally screened by tussock.

An observer states that, when the harrier wheels round, and appears about to settle, the larks often gather together in numbers with a chirping sound, and move about restlessly, sometimes with a short flight, watching the movements of the enemy. Some larks rescued from the attacks of hawks have been so completely prostrated by fear as to be quite incapable of flight, and, on being tossed gently into the air, have fallen helplessly to the ground.

Nest of Ground Lark.

A theory has been put forward maintaining that, when the lark takes a dust bath, which it does fairly frequently, it is attempting to rid itself of the persecution of some parasitic vermin. It has a habit of keeping its insect prey in its beak for a long time before it is devoured or carried off to the nest.

White larks, which are exceptionally beautiful little creatures, have been seen in Canterbury, generally in scattered companies.

The Antipodes Lark.

Anthus steindachneri.

Like the last, but the lower surface tinged with fulvous, and the wing rather shorter. Length of the wing, 3.25 in.; of the tarsus, 1 in. Antipodes Island. The crest of the sternum is much reduced and the bird flies very badly.

Family Meliphagidae.

Bill rather long and usually curved. Tongue protractile, furnished at the tip with a bunch of stiff fibres. Tarsi rather short, scutellated; the scutellæ sometimes fused together.

Key to the Genera.

1.	Tufts of white feathers on the throat.	Prosthemadera.
	Tufts of white feathers behind the ears.	Pogonornis.
	No tufts of white feathers.	2
2.	White feathers round the eye.	Zosterops.
	No white feathers round the eye.	Anthornis.

Key to the sub-families.

First primary nearly as long as the second.	Zosteropinæ.
First primary one half of the second.	Meliphaginæ.

Genus Zosterops.

Bill nearly straight, as long as the head. First primary very short, the second long, the tip of the wing formed by the third and fourth. Tail emarginate. A ring of white feathers round the eye. Africa, India, Malay Archipelago, Polynesia, Australia, and New Zealand.

The White-eye.—Tau-hou.

Zosterops caerulescens.

Summer plumage—Head and over tail bright olive; back dark grey; throat yellow; abdomen white, the flanks light chestnut. Winter plumage—Chin and throat light grey. Eye reddish brown. Length of the wing, 2.3 in.; of tarsus, 0.6in. Egg—Pale blue; length, 0.7 in. Both islands, Chatham Islands, Auckland Islands, Campbell Islands, and Australia.

The white-eye is an Australian colonist. It first came to New Zealand about the year 1856, and it liked this country so well that it has stopped with us ever since, increasing rapidly in numbers. It is both welcome and unwelcome; that is to say, it does harm and good. By feeding on the American blight, which attacks apple-trees, it has made its presence very beneficial. In recognition of its efforts in this direction, it has been awarded the title of "blight-bird." But, when the fruit is ripe, it undoes a great deal of the good it has effected, by making depredations on the cherries and plums.

Its nest is made up of moss and grass, and although fragile in appearance, it is in reality of great strength. It is firmly secured to a forked twig by silky threads of spiders' webs. It is finished on the outside, round the bottom, with braces of green leaves of grass, crossed and recrossed, which adds much to the strength and stiffness of the fabric. Three or four eggs are laid in the nest.

Before 1856, the white-eye was unknown to either Europeans or Maoris in New Zealand, and, when it came here, in that year, the latter named it "tau-hou," which means "stranger." By 1861 it had spread all over the South Island, and the southern parts of the North Island, though it did not reach Auckland until 1865. It has also appeared in the Chatham Islands, the Snares, the Auckland Islands, and the Campbell Islands, and has become naturalised there. Its presence here shows that the passage across the Tasman Sea is possible even for some small land-birds. The distance, as the crow flies, is about a thousand miles; and it must take a bird, flying at ordinary speed, from twenty-four to thirty-six hours to accomplish the journey.

Genus Prosthemadera.

Bill rather longer than the head. Fifth and sixth quills nearly equal and longest; the third to the sixth sinuated on their outer edges. Tail long and rounded. New Zealand only.

The Tui.—Tui or Koko.

Prosthemadera novae-zealandiae.

Bluish or greenish black, with white streaks on the back of the neck, and a white spot on each wing. Throat ornamented with two tufts of white curly feathers. Eye dark brown. Length of the wing, 6.5 in.; of the tarsus, 1.5 in. The sexes are alike. The young are slaty black with a light patch on the breast. Egg—White or pale pink spotted with rufous brown, principally towards the larger end; length, 1.3 in. Both islands, Chatham Islands, Auckland Islands.

Although the tui is called the "parson bird," on account of the little tuft of white feathers that sticks out from its throat, and contrasts with its sombre plumage, it is not very sedate or staid in its bearing. Its flight, indeed, is distinguished by gaiety, and sometimes it sends forth a wild burst of joyful notes. Mr. Hursthouse, in his *New Zealand*, has given a brief but admirable description of this bird. "It is larger," he says, "than the

Tui: old and young.   *(Buller.)*

black-bird, and more elegant in shape. Its plumage is lustrous black, irradiated with green hues, and pencilled with silver grey, and it displays a white throat-tuft for its clerical bands. It can sing, but seldom will; and it preserves its voice for mocking others. Darting through some low scrub to the topmost twig of the tallest tree, it commences roaring forth a variety of strange notes, with such changes of voice and volume of tone as to claim the instant attention of the forest. Caught and caged, it is still the merry ventriloquist, and mocks cocks and cats, and attempts the baby. To add to its merits, it becomes very fine eating in the season of poroporo berries."

Sometimes the tui assumes an antagonistic attitude towards other birds, and it has been seen chasing the kingfisher, the bell-bird, and other enemies from the trees in which its young have been placed.

The nest, which is large, is constructed of slender sprays intermixed with moss, and with the down of the tree-fern, and lined with the fine bents of the poa grass. The following dimensions of a nest have been supplied; across the top, from the outside of one wall to the outside of the other, 9 inches; diameter

Nest of Tui.

of cavity, 3 inches, 6 lines; depth, 2 inches. There are generally four eggs. The nest containing the young is sometimes stained a deep purple, from the juice of the konini berries. It is stated that, on one occasion, some young birds which were unable to fly, became alarmed and fluttered out of the nest to the ground, about twelve feet below. The next day they were found safely ensconsed in the nest, looking quite happy. They must have been replaced by the parent birds.

Tui-pie was a favourite dish with the early settlers. The birds are often kept in confinement, and at one time many of them were sent to Australia and other countries. The Maoris had no fewer than seven methods of catching tuis, which were sometimes treated as pets. With one method, birds were induced by imitation calls to alight on a perch, and were then knocked down by means of a long flexible stick. An experienced and skilful fowler could catch as many as a hundred in one day. They were also speared and snared by means of a specially prepared apparatus. In the frosty weather they were sought at night, and were captured without much trouble, as their claws were contracted by the cold, and they could not fly.

A writer in an early number of *The Ibis* refers to the peculiar manner in which the tui mounts into the air in fine weather, parties of about half a dozen "turning, twisting, throwing somersaults, dropping from a height with expanded wings and tail, and performing other antics, till, as if guided by some preconcerted signal, they suddenly dive into the forest and are lost to view."

The song of the tui, as represented in the Maori language, has been written by Sir George Grey, and recorded in his work entitled *The Poetry of the New Zealanders*. Sir George Fenwick, of Dunedin, in describing a trip to Milford Sound in the early part of 1903, says that he heard the charming melody of this bird everywhere he went, that it met the ear at almost every step in the Clinton and Arthur Valleys, and was also heard in the trips on the Lakes and Milford Sound. The variation of the note in different localities was well exemplified in the districts he traversed. He says that the bird seems to have largely discarded the somewhat harsh note with which it usually ends its song in other parts of New Zealand, and has substituted, very often as its only song, what might be termed a single staccato note, which it repeats from four to six times. If the note F on the treble clef above the middle C is struck on the piano and then whistled either four or six times, an imitation of this particular note of the tui will be produced

"More often than otherwise," he says, "it is satisfied with the repetition of this note only, but it frequently adds another note, which may be stated as

The notes are uttered very rapidly, and are very melodious and pleasing.''

Genus Pogonornis.

Bill shorter than the head. Fourth and fifth quills of the wing equal and longest, the first long. Tail rounded, each feather pointed. New Zealand only.

The Stitch-Bird.—HIHI or TIHE.

Pogonornis cincta.

Male—Head and neck black, with a tuft of white feathers behind each ear. Breast and some of the wing-coverts bright yellow. A white band on the wings. Abdomen brownish white. Eye black. Length of the wing, 4 in.; of the tarsus, 1.2 in. Female and young—Brown, with a white band on the wings. Egg—White, thickly spotted all over with rufous; length, 0.75 in. North Island only.

The stitch-birds have become very rare on the mainland, but they may still be seen on Little Barrier Island. They are peripatetic little birds. With their heads carried proudly, their wings drooped, and their tails spread and raised, they flit about in the trees, always on the move. As each movement is made, they give a peculiar whistle.

Of all the birds I saw on the Little Barrier Island when I visited that sanctuary in February, 1907, none interested me more than the stitch-bird. It is one of the rarest of all the rare birds of New Zealand. In the palmiest days of bird-life in the dominion, about thirty years ago, it was often seen, but I do not believe that it can now be found anywhere except on the Little Barrier. It is not a very shy bird, but it frequents inaccessible places in the densely-wooded mountain gorges. Its great rarity, the interest taken in it by naturalists in all parts of the world, and the large number of quests that have been made to obtain specimens of its skin for museums and collectors in the Old Country, made me determined to see something of it if I could. My opportunity came only two days before my departure. I went with the members of Mr. R. H. Shakespear's family,

(*Buller.*)

Stitch Bird: male (below) and female (above).

who reside on the island, for a day's walk to the top of one of the mountain peaks. At noon, when we were making our way along a track through the forest on the mountain side, a female stitch-bird, which had evidently come from the heights, fluttered about excitedly in the boughs above our heads. The cry sounded like the words "steech, steech," and when this was imitated she came closer and flew amongst some saplings a few yards away. With her tail erect and almost at right-angles with her body, and her wings drooping, she ran up and down the boughs and turned round frequently, as if she was the embodiment of motion. She hardly ceased to give the cry, and there were few moments that she took her eyes off us. Later on, she was joined by a few companions, all of them females. I was disappointed at not seeing a male stitch-bird, which is an exceptionally handsome bird, but deemed myself fortunate in having had a passing acquaintance with the female.

The interest taken in this bird by naturalists is shown by the fact that Mr. Reischek, the Austrian collector, visited the island on several occasions to see it. In 1880, he camped on the island for three months, without meeting with any success. Two years later, he sent his assistant, who remained for some time, and succeeded in shooting only one pair. Reischek visited the island again, with a determination to live there until he had seen a stitch-bird. After five weeks' continuous search, during which he laboriously traversed nearly all parts of that rugged and precipitous little island, he was at last rewarded by hearing the stitch-bird's whistle. He was not able to get near enough to see the bird on that occasion. With the assistance of a friend who accompanied him, he cut tracks to the tops of some of the mountains, but they could not see the object of their search, although they often heard its call. Reischek then removed his headquarters towards the centre of the island. Shortly afterwards his eyes were delighted with the sight of the stitch-bird. He was so interested and excited in watching this rare and beautiful bird's movements that it disappeared before he had time to use his gun. He continued his quest for three more weeks before he was able to shoot any stitch-birds. He had discovered their haunt. It is a deep ravine near the top of the range, where rocks form steep precipices. It took him two days tramping, climbing, and scaling cliffs to get back to the landing-place on the shore; but he felt that he had been well rewarded

H

for his trouble, as he had visited the hihi, as the Maoris call it, in its last home.*

Genus Anthornis.

Bill equal to the head, curved. First quill rather long, the second abruptly narrowed near the tip in the adult, acutely pointed in the young; fifth the longest. Tail emarginate. New Zealand only.

The Bell-Bird.—KORIMAKO (North Island) and MAKOMAKO (South Island).

Anthornis melanura.

Male—Yellowish olivaceous, the head tinged with steel black. Wings and tail brownish black. Eye blood red. Length of the wing, 3.5 in.; of the tarsus, 1 in. Female and young—Brownish olivaceous; wings and tail, brown; a white line from the bill towards the sides of the neck. The female is rather smaller than the male. Egg—Pinkish white, with irregular markings of reddish brown, principally near the larger end. Both islands and the Auckland Islands.

Over a hundred years ago Sir Joseph Banks described the delight he experienced when this bird's splendid song first fell upon his ears. It was in Queen Charlotte Sound, when Captain Cook's vessel was about a quarter of a mile from shore. "And in the morning," he says, "we were awakened by the singing of the birds. The number was incredible, and they seemed to strain their throats in emulation of each other. This wild melody was infinitely superior to any that we have ever heard, of the same kind: it seemed to me like small bells, most exquisitely tuned; and perhaps the distance and the water between might be no small advantage to the sound."

The bell-bird spends a great deal of its time in the summer sipping nectar from the blooms of the *Phormium tenax,* and other plants. Very often it is seen in company with its friend the tui. Mr. W W Smith, who has watched these two birds closely, says

---

* By J. Drummond, in the Sydney *Morning Herald,* June 1st, 1907.

that there is no picture in Nature more beautiful than the sight of them clinging and swinging in grotesque postures in the sunshine on the brilliant blooms of the rata, flying every few minutes to some bough, and uttering their rich songs. The bell-birds and the tuis come to some of the native trees, and remain there as long as the blooms support them. They then disperse among the warmer valleys of the bush, and subsist, during the wet winter months, chiefly upon insects, until the return of the spring, when the melliferous blooms of the kowhai again supply them with the necessary food.

The decrease in the numbers of the bell-birds is attributed to bush fires, especially in the North Island, and to the depredations of cats, rats, and honey bees. In many districts bell-birds have entirely disappeared since the bees were introduced. A theory has been advanced that the honey-eaters are stung to death when attacking and eating the bees. Mr. Smith, after close study and observation, has come to the conclusion that the honey-bee theory, in its bearing on the disappearance of any of our native birds, is an utter fallacy. He admits that the bees rob the honey-eaters of a little nectar during the season, but he says he has never seen the bees in such numbers as to affect the honey-eaters. He has watched the birds and the insects daily for years regaling themselves together on the yellow kowhai blossom, but he has never seen the honey-eaters attack the bees. He has also watched them feeding together for hours on the peach blossoms without the least signs of molestation. Inquiries show that his conclusions are probably correct. It is probable that cats and rats are the birds' great enemies.

In September the male and female bell-bird begin to build a nest together. It is made of small branches and moss, lined with feathers, and is placed in thick branches of trees, from 20 to 50 feet above the ground. Nests have also been found in hollow trees. At the end of September, or early in October, four or five eggs are laid. The two birds hatch them together, and both parents feed the young. When leaving the nest, the male looks after the family until they are able to take care of themselves. A male bird has been seen to knock a young one from a branch when it would not listen to his call announcing approaching danger.

Silver-eye or White-eye.

(*Buller.*)

Bell-bird: male and female.

The bell-bird is both bold and tame. It eats berries, as well as insects and honey, but its special weakness is the honey in the flower of the *Phormium tenax*, the delicious liquid being obtained by inserting the bushy tongue into the calyx.

It is interesting to note a statement that the planting of acacia trees greatly helps the bell-birds through the winter months and the early spring, as the flowers of these trees supply a food of which the bell-bird cannot be deprived by blackbirds and other introduced species. Mr. Potts planted a few of several kinds of acacia specially for the bell-bird, and he was rewarded by a constant melody through subsequent winters.

Nest of Bell-bird.

A description of the nests states that it is rather flat, with a well-formed cup, loosely yet strongly built of sprays, grass, and moss, and well lined with feathers. From wall to wall across the top, it measures about five inches. The diameter of the cup is two inches and three-quarters, and the depth inside is two inches. Attention has been called by Mr. Potts to the peculiarity of the colouring in the lining feathers, as betokening the bell-bird's love of harmony. Some of the nests are lined with red feathers from kakas, green ones from parrakeets, black ones from Norfolk turkeys, buff from Cochin fowls, speckled from pintadoes, and white from geese. Nests are often found in much frequented places, such as in a shrub on a public road.

The Maoris caught the bell-bird by means of a long spear, called a "here." It was made from a carefully selected, straight-grained piece of wood, and was sometimes thirty or forty feet long. It was fitted with a barb, made of bone. One end of the barb was sharpened by scraping, and was serrated, so that it would hold the bird when the latter was struck. Other contrivances were also used.

(Voy. Erebus and Terror.)
Chatham Island Bell-bird.

A very encouraging note is sounded by Mr. W W. Smith in some interesting information he supplies in respect to this charming songster of the forest. He states that, for several years, he has frequently visited the Alford Forest and Mount Somers bush, in Canterbury, and Peel Forest and the Albury bush, in

South Canterbury, and on each visit he has found that the species is increasing. When he stayed at Windwood, Mount Somers, he found both adult and young inhabiting the warm wooded valleys of the Gawler Downs in good numbers. At Albury, bell-birds were seen in considerable numbers. Twice he was on the top of Rocky Peninsula at daybreak, listening to the waking melody of the birds echoing across the gorge from the opposite bush. On both occasions the morning was serene and beautiful, and the mingling of the songs of numerous birds with the soft murmur of the river far below presented to him one of those enchanting scenes in bird life daily realised by ornithologists out in the open in New Zealand.

Nest of Chatham Island Bell-bird.

The Chatham Island Bell-bird.

Anthornis melanocephala.

Like the last species but larger. In the male the head is steel black, and the neck, breast, and upper tail coverts are tinged with black. Egg—Pink, sparingly spotted and blotched at the larger end with chestnut; length, 1 in. Length of the wing, 4.25in.; of the tarsus, 1.5 in. Chatham Islands.

The Chatham Islands have a bell-bird of their own. Its note is said to be much richer and fuller than that of the New Zealand species. It begins to breed in October. The nest is composed of grass and feathers, and is coarsely constructed. The female, as a rule, lays three eggs.

Sub-order Clamatores.

With a simple vocal apparatus.

Family Xenicidae.

Bill not notched at the end; no mouth bristles. Tail very short, of ten feathers. Tarsi booted. New Zealand only.

Key to the Genera.

1. Tarsi one inch in length.	Xenicus.
Tarsi less than one inch in length.	2
2. Bill nearly as long as the tarsus.	Traversia.
Bill much shorter than the tarsus.	Acanthidositta.

Genus Xenicus.

Bill moderate. Wing with the third to the fifth quills nearly equal and longest; the second rather shorter than the seventh. Legs long, the feet rather strong. Tail short.

Key to the Species.

Breast grey.	X. longipes.
Breast tawny.	X. gilviventris.

Green Wren.—MATUHI.

Xenicus longipes.

Above green, top of the head purplish brown. Chin white, breast grey, flanks and vent yellowish green. A broad white streak over the eye. Eye dark brown. Female—head and upper back, brown; lower back, olive green; below, grey; the flanks and vent, greenish yellow. Stripe over the eye as in the male. Length of the wing, 2.25; of the tarsus, 1 in. Egg—White with irregular blotches at the thick end. Both Islands, in sub-alpine forests.

The New Zealand green wren, like "Jenny Wren, the Fair," is little but good. It is among the smallest of the birds of this dominion, and it may often be seen creeping among the lichens and mosses that decorate the stem and branches of the forest

Bush Wren: male.   Bush Wren: female.

Rock Wren.   Green Wren.

(Buller.)

trees. It is lively, confident, and restless, and has a sharp note.
Most of its time, apparently, is spent in minute investigations
into the lichens and mosses. Mr. Potts has also seen it specially
busily engaged where the level velvety surface of the ground

has been disturbed and upturned by the strong claws of the wood hens.

The nest is admirably hidden, generally amidst a quantity of moss. A typical one was found beneath the moss-covered roots of a ribbon-wood tree. It was pouch-shaped, and had an opening near the top. The sides were straightened with fern root carefully interlaced and beautifully interwoven. It was so well concealed that the entrance could hardly be seen.

Nest of Green Wren.

The Rock Wren.

Xenicus gilviventris.

Above, olive green; the head and neck, brownish. Below, fawn colour; the flanks and vent tinged with yellow. A white line over the eye. Eye dark brown. In the female the upper part of the back is brown. Length of the wing, 2 in.; of the tarsus, 1 in. Egg—White; length, 0.7 in. South Island.

This little bird is found only in the mountain regions of the South Island among stunted vegetation, and is rarely seen. Several dead specimens have been picked up on the glaciers. It haunts chiefly the rocky debris at the foot of mountain peaks, and runs under the stones like a lizard. It feeds upon insects. and forms its nest among the rocks.

Stephen's Island Wren.

(*Ibis.*)

Genus Traversia.

Bill large and stout, nearly as long as the tarsus. Tarsus short, about as long as the hind toe with claw. Wing weak; plumage soft.

The Stephen's Island Wren.

Traversia lyalli.

Male—Above, dark brownish olive-yellow. A narrow yellow line over the eye. Wings and tail brown, the wing coverts like the back. Chin, throat, and breast yellow, the flanks and abdomen pale brown. Female—Above, brown; below, buffy grey. Length of the wing, 2 in.; of the tarsus, 0.75 in. Stephen's Island, in Cook Strait.

This bird, it is believed, is now extinct, having been killed off by the cats. It is said to have been semi-nocturnal in its habits, frequenting the scrub on the island.

Nest of Bush Wren.

Genus Acanthidositta.

Bill slender. Wing with the third to fifth quills nearly equal and longest, the second rather longer than the seventh. Tarsi long and slender.

The Bush Wren.

TITITI-POUNAMU.

Acanthidositta chloris.

Male—Above, green. Wing feathers, dark brown, edged with green. A white line over the eye. Below, white tinged with yellow on the flanks. Tail black with a yellowish white tip. Eye dark brown. Female — Brownish white, streaked with dark brown above; below, white. Length of the wing, 1.5 in.; of the tarsus, 0.75 in. Egg—White; 0.6 in. Both Islands.

The bush wren, or rifleman, inhabits generally the sub-alpine forests of both Islands as well as the Great Barrier Island and the Little Barrier; but very little is known of its habits. It forms its nest in hollow trees, and it is stated that a pair of these little birds once took up their residence in the skull of a horse.

ORDER HALCYONES.

Legs and feet weak, the outer toe united to the middle one for half its length.

Family Alcedinidae.

Bill, long, strong, and pointed, not serrated. Wings and tail rounded.

Found over the whole world.

Genus Halcyon.

Bill compressed. Second feather of the wing longest, the secondaries nearly as long as the primaries. Tail with twelve feathers, longer than the bill; rounded at the end. Eye, black. Africa, Southern Asia, and Australia.

The Kingfisher —KOTARE.

Halcyon vagans.

Top of the head and upper part of the back, green; lower back, bright greenish blue. Wings and tail dark blue. Under parts and a band round the neck, buffy white. Generally there are some brown markings on the breast. Length of the wing, 4 in.; of the tarsus, 0.5 in. The young are brown above, with a few white feathers on the back of the neck. Wing coverts tipped with yellow. Some of the breast feathers broadly margined with brown. Egg—White; length 1.2 in. Both Islands, Norfolk Island, and Lord Howe Island.

Sitting on a bough that stretches out over a stream, watching the waters beneath, and wrapped up, apparently, in a brown study, the kingfisher often spends hours together. He seems to have no more serious business on hand than to display his gaudy coat of many colours in the sun. Suddenly, however, all outward signs of mental inactivity vanish. A few rapid strokes of the wings, and, with his spear-like bill thrust well forward, and his wings tucked in, he darts through the air, a living flash of light, touches the surface of the water, and secures the fish for which he has waited so long. A sudden twist, a few more rapid strokes of the wings, and he returns to land with his prey, knocks it on a stone or other hard substance until it is dead, or squeezes it to death between his powerful mandibles, and then swallows it whole, most likely head first.

The European kingfisher eats nothing but fish. It is their proficiency in this direction that has earned for the members of

Kingfishers: male and female.   (*Sharpe.*)

the family the title of "kingfisher," to which they have good claims. But the New Zealander does not live by fish alone. He has an insatiable appetite, and will eat mice, bees, beetles and other insects, crustacea, and even young birds and chickens. The omnivorous little rascal was accused, some years ago, of catching trout in the Acclimatisation Society's Gardens in Christchurch, until the artificial races were netted in to keep him off, and also of killing Californian quail in the Auckland Gardens.

Our kingfisher, like other members of the family, is a grave, sedate, mournful, and usually silent bird. It is a worthy representative of "the mournful race," as Dryden has called them. Even when not engaged in fishing excursions, it often sits motionless, like Patience on a monument, without uttering a sound. It never delights to flit about in the trees, as if happy in existence. It has no real song; and its cries are monotonous and harsh, and without any sweet music.

In spite of moroseness and gluttony, the kingfisher has several good qualities, which appeal to our leniency.

It is persevering. This fact was proved beyond doubt when a pair tried to build their nest in the back of a plastered sod chimney attached to an empty cottage. It should be stated, first of all, that the nest is like a robber's cave. It is generally tunnelled into a bank, wall, or tree. The tunnel is about sixteen inches long, and two inches in diameter at the entrance. The floor of the tunnel rises from the entrance. It leads to an inner chamber about seven inches long and five and a half inches wide, while the height from the floor to the roof is four inches. This is the egg chamber. It is hollowed out slightly below the floor of the tunnel, so that neither the eggs nor the young, when they are first hatched, can roll out and meet with accidents. The young remain within the walls of this comfortable little nursery until they are ready to fly well.

The incident in connection with the cottage, which is related by Mr. Potts, happened a good many years ago. It is sufficient to say it was the month of October. On the nineteenth, the kingfishers, a male and female, started working at the chimney. They completed the tunnel; but, shortly after they had commenced on the egg chamber, they abandoned the nest, probably because they found that the wall of the chimney did not give what they deemed to be sufficient depth for the safety of their offspring. On November 3rd, they were seen again, hard at work on a new nest, in front of the cottage, this time between the door and the window. This also was deserted, probably for the same reason. On November 14th, they had chosen another site on the southern wall of the cottage. There they were seen darting upwards from a convenient rail five or six times a minute, and digging out the material with their bills, which were used as picks. It was hard work; and it was all in vain, as this site also was unsuitable, and had to be abandoned. Three more attempts were made; three more failures were recorded. Finally, on November 26th, thirty-eight days after the project had been started, success crowned their efforts, and they reaped the reward of perseverance and industry. It was the seventh attempt, and the spot proved to be on the very wall of the chimney that had been the scene of the first attack. On December 4th, the nest was found to contain two eggs. Later on there were five, and, on December 24th, the family was increased by five young kingfishers.

Sometimes the construction of a nest occupies several weeks. The labour is divided fairly equally between the male and the female. When a pair were timed, it was found that, if the female worked harder on one day, the male made up for it the next day. When a site is finally selected, and the co-operative works are fairly started, the birds do not leave the spot, but one keeps watch while the other is working, or seeking food. If an alarm is given, it is quickly answered, perhaps from a distance of half a mile. A bird will stay in the tunnel sometimes for three minutes. It will emerge head first, having turned round inside, and its place will be taken by the mate.

It is recorded by Mr. Potts in connection with one pair that, when the female flew off to feed, the male remained to watch just below the hole. As soon as his mate returned, in about twenty minutes, he recommenced his work. They darted upwards from their perches into the hole, always correctly judging the distance, and, at the moment of entering, uttered a short cry of two notes, like ''chi-rit.'' Once the female darted to the hole, flew back, perhaps out of timidity, then sought the male, who bent down from his perch and caressed her with his beak.

They will tolerate no intrusions into their home, and resent trespassers even in the vicinity. The female has often been seen to meet a person two or three hundred yards from her young, dash at the intruder, return to the place where the young were perched, and repeat the attack several times. A kingfisher has been known to attack and drive back a dog, and, on some occasions, Mr. Potts states, it will make one of its famous lightning darts into a group of pigeons or other birds, merely, apparently, for the sake of standing by and enjoying their terror. While sheep and cattle have been allowed to graze close to a nest, a cat, a dog, or a human being has been determinedly attacked.

There is not much more to add in regard to our kingfisher. It is a useful bird, and eats up many injurious insects. One of its characteristics is a great power of vision, and it has performed some remarkable feats in sighting insects and bearing down on them. It is a creature of habit, returning time after time to the same nest or perching place. The New Zealand species seems to be rather more sociable than the European bird. It lays from five to seven eggs. Breeding takes place twice a year.

The family, which is found all over the world, is celebrated in poetry and legendary lore. The order still retains the name "Halcyon," given by the ancients, who believed that the seven days before the winter solstice and the seven days after were the halcyon days, when the sea remained calm, so that the bird could build its nest, which was believed to be placed on the ocean. After death, its body was awarded attributes, such as the power of lulling the tempests and giving beauty and prosperity. To this day there is a belief in some parts of Europe that, if a dead kingfisher is suspended by a thread, it will turn its beak towards the wind and show from which direction it is blowing.

ORDER CUCULI.

Family Cuculidae.

Bill, moderate, or slender, slightly curved. Wings and tail, long. Legs and feet, small; two toes in front and two behind.

Key to the Genera.

Colour, metallic green.   Chalcococcyx.
Colour, brown.   Urodynamis.

Genus *Chalcococcyx.*

Bill moderate, the nostrils round. Third quill the longest; under surface of the quills with a single oblique pale bar. Tail rounded, of ten feathers. India and the Malayan Archipelago, to Australia and New Zealand.

The Shining Cuckoo.—PIPIWHARAUROA.

Chalcococcyx lucidus.

Above, metallic bronzy green; below, white, barred with bronzy brown. Outer tail feather barred with white, the second with rufous. Forehead, freckled with white. Eye, black. Length of the wing, 4 in.; of the tarsus, 0.65 in. The sexes are alike. The young are duller in colour, and the bands on the lower surface are more numerous and not so distinct. Egg—Brownish olive; length, 0.75 in. Both Islands, Chatham Islands, Northern Australia.

Both our cuckoos sustain the reputation of many members of the Cuculidæ family as impudent parasites, who usurp the nests of other birds, and leave them to hatch and bring up the young cuckoos. One of our birds is called the shining, or bronze, cuckoo, and the other the long-tailed cuckoo, and both are summer visitors and notable migrants.

In New Zealand, the little grey warbler is generally selected by the cuckoos as a victim for their parasitical habits. Instances are recorded, however, where the nests of other birds have been appropriated. Mr. G. M. Thomson, of Dunedin, states that he was once shown a house-sparrow's nest, which had been built in a large bramble bush, and contained four eggs, three belonging to the sparrow and one to the shining cuckoo. In Otago the grey warbler, the South Island tom-tit, and the white-eye are generally made the foster-parents. In *Gatherings of a Naturalist in Australasia,* Dr. Bennett states that a white-striped fly-catcher was shot at Ryde, near Sydney, in the act of feeding a solitary young bird in its nest, which, when examined, was found to be a chick of the shining cuckoo. He says that it was ludicrous to observe the large and, apparently, well-fed bird filling up with its corpulent body the entire nest, and receiving daily the sustenance intended for the young fly-catchers.

The proceedings of the cuckoos in regard to their parasitical habits are a mystery. The actual presence of the parasite's egg in the warbler's nest is one of the strangest features of the affair. The nest is usually the shape of a soda-water bottle; it is eight inches long, and about four inches in diameter at its

Shining Cuckoos.   *(Buller.)*

widest part. The entrance, which is at the side, is almost an inch and a half across, and about one inch perpendicularly. The upper portion of the nest somewhat overhangs the aperture, forming a kind of hood. It would be almost impossible for a cuckoo to enter the nest and lay the egg in it. The supposition is that the egg is laid on the ground, and is carried by the cuckoo hen and deposited in the nest.

From observations made by many naturalists, it seems that the cuckoo goes about its business with a great deal of forethought and cunning. In order that the victim will not be likely to abandon the cuckoo egg, it is nearly always placed in a nest that contains other eggs. But, by-and-by, the young cuckoo becomes the sole occupant. In New Zealand, the little intruder probably follows the tactics adopted by these birds in England, where the young cuckoo lifts the other young birds on its back, and tumbles them overboard. Mr. Potts says that, in all his experience, he has known only of one egg of a shining cuckoo to be broken in the nest. This is a remarkable fact, considering the thinness and fragility of the shell, and the distance from the entrance of the nest to the bottom.

A writer in *The Zoologist*, Mr. A. H. Meikelejohn, claims to have the unique experience of seeing an English cuckoo in the act of endeavouring to place its egg in the nest of a pair of robins. The incident, he reports, took place between the villages of Hamstreet and Woodchurch, Kent, England. A hen cuckoo carried her egg in her throat to the nest of the robins, who attacked her with great fury. Again and again the little birds struck and buffeted her, and on two occasions one of the robins seized hold of her by the back of the neck, and held on for a few seconds with all the fierce tenacity of a bull-dog. Ultimately, she suddenly darted amongst the grass and disappeared.

The young cuckoo is fairly well feathered before it leaves the nest. It cries almost incessantly, and is fed by the foster-parents till it can fly well. When attended to by the warblers it moves its head with a lateral motion, in a watchful manner, like a hawk. As the warbler approaches, it displays an eager interest by slightly opening and fluttering its wings, and hopping forward to meet its feeder. At this time the cries are repeated more rapidly. Then the insect, brought by the warbler, is swallowed, and the monotonous cries are resumed. Although the young cuckoo may be surrounded by the curious of many species of birds, Mr. Potts adds, it takes no heed of them. When it becomes self-supporting, and obtains food for itself, it shows a great fondness for the larvæ of the kowhai moth.

Our cuckoos do not possess the "cuckoo" note which has given the bird its name. The notes of members of the Cuculi order vary considerably, and Dr. R. Bowdler Sharpe states that only one other species, which is found in South Africa, has the same cry as the European species. The cry of the shining cuckoo is described as being an exceedingly pleasant one, consisting of a number of silvery notes. It has also been reported that this bird is endowed with some kind of ventriloquism, which makes it seem to be far away when it is really quite close.

A few brief notes from the diary of Mr. W. W. Smith record the stages of a domestic tragedy in bird-land, brought about by the cuckoo's strange habits. He once found the nest of a grey warbler, and the bird flew out as he approached. There were five eggs in it, and they were all intact. A few days later, on September 21st, he returned, and the eggs were still unhatched. Other entries in the diary are:—"September 24th—Two young ones hatched; one egg, lying on the ground outside the nest, contained a chick, cold and dead. September 25th—Three young ones in nest; large egg unhatched. September 26th—Large egg hatched—chick of a shining cuckoo; very clumsy in nest, lying on the top of the three young warblers. September 30th— Found one dead chick, lying on the ground; two young warblers still alive; young cuckoo growing rapidly, being now nearly large enough to fill the nest itself, beak and legs fairly well developed. November 2nd—One of the young warblers lying dead in nest, the other alive; young cuckoo has its eyes open; signs of feathers on the neck and wings, but in parts of the body perfectly bare. November 5th—Visited nest again; young cuckoo thrust out its head to receive food when I approached; lifted the surviving young warbler out of the nest and found it very feeble; the young cuckoo lying with its head at the opening of the nest, had taken full possession; its lifeless companion was lying underneath, having, apparently, died from starvation. November 8th—Found the young cuckoo almost ready to leave the cradle; brought both nest and bird home. November 10th— Thriving well, being fed on small worms, crabs, flies, spiders, and very small pieces of lean meat. November 15th—Has now come out of the nest; eats largely three times a day, but does not care for meat; increasing rapidly in size. November 20th—Nearly feathered; placed it in the cage, but it looked sickly. November 21st—Young cuckoo died.

It was by observing the habits of the shining cuckoo that naturalists found that New Zealand participated in the great southern migrations. When Mr. Colenso stated, in 1842, that the bird was migratory, the furthest distance across the sea that migratory birds had been known to fly was from Norway to Scotland, and across the eastern Mediterranean from Egypt to the Greek islands, in each case a distance of about 300 miles, involving about eleven hours of continuous flying. When it was asserted that the shining cuckoo traversed more than three times that distance of ocean, from New Caledonia to New Zealand and the Chatham Islands, it was thought that naturalists here had made a mistake. Dr. Wallace, in his *Geographical Distribution of Animals,* which was published thirty-four years after Mr. Colenso's statement, says that it is extremely improbable, and he adds that ''in a country which has still such wide tracts of unsettled land, it is very possible that the birds in question may only move from one part of the islands to the other.'' It is now fully acknowledged, however, that these birds do migrate, and that they are among the most notable migratory birds in the world.

The shining cuckoo appears in the northern parts of New Zealand regularly in the latter half of September; and early in October it is found in Wellington and in the South Island. It breeds in New Zealand. All the old birds leave the southern portions of the country during the first and second weeks in January, but they do not leave the north until the end of January, or perhaps later. Some, at least, of the young birds leave considerably later than their parents, as they have been shot in the South Island in April. The times of appearance and departure of the old birds are wonderfully regular in both islands. In the Chatham Islands the birds come and go at about the same dates as in New Zealand. Here we have distinct evidence that the birds travel from the North to the South, and then back again to the North. They have not been seen to leave the Islands, but it is impossible that they could remain during the winter and yet escape the eyes and snares of the Maoris, for there are no wide tracts of unsettled land for them to go to either in New Zealand or in the Chatham Islands. Mr. Potts says that he once saw the arrival of a shining cuckoo at the Chatham Islands. It was so tired when it landed that it allowed him to pick it up in his hands, although in ordinary circumstances it is a very shy bird. Dr. E. P Ramsay has in his collection a specimen which was taken at sea between New Zealand and Lord Howe Island.

Outside New Zealand there is little information. The species is found in Norfolk Island, where it also breeds, but elsewhere it has been obtained only at Cape York, in North Queensland, where it is very rare; and it is probable that its winter home is in New Guinea. Mr. E. L. Layard has stated that it occurred in New Caledonia, but, according to the authorities of the British Museum, his specimens belong to the allied species, *C. plagosus,* which migrates from north to south in Australia, but does not come to New Zealand.

Genus *Urodynamis.*

Bill strong, arched. Fourth quill the longest, the second shorter than the fifth. Tail very long and wedge-shaped. New Zealand and Polynesia to the Solomon Islands.

The Long-tailed Cuckoo.—KOEKOEA.

Urodynamis taitensis.

Above, brown, banded, and streaked with rufous. Below, white with longitudinal streaks of dark brown. Eye, reddish brown. Length of the wing, 7.75 in.; of the tarsus, 1.4 in. The sexes are alike. The young have the upper surface brown, spotted with fulvous white; the lower surface rufous white, streaked with dark brown. Eye, reddish brown. Egg—Brownish olive, sometimes clouded with brownish grey; length, 0.7 in. Range, the same as for the genus.

The long-tailed cuckoo arrives in New Zealand at the end of October or beginning of November, and leaves in January or February, but its movements are not so easily traced as those of the shining cuckoo. The young birds of this species also linger longer than their parents, and are occasionally seen as late as the first week in April. These birds retain their young spotted plumage much longer than the young of the shining cuckoo do, but no specimen showing the change into that of the adult has as yet been shot in New Zealand, and neither old nor young have been recorded from the Chatham Islands.

(*Buller.*)

Long-tailed Cuckoo: old and young.

Mr. Layard says that this bird is very rare in New Caledonia. Ie obtained only four specimens, all of which were purchased in he streets. The first was on March 23rd, 1879, the second on Iarch 15th, 1881, and the other two on April 15th, 1881. As the irds were in their immature plumage, he thinks that they were orn in the island.

Mr. C. W. Woodford, Resident Commissioner of the British olomon Islands, states that he obtained immature males of the ong-tailed cuckoo in April and May, 1887, and that he has seen he bird several times in the Solomon Islands during the past hree years, the last time being in May, 1900. He is of opinion that the bird is a migrant, but cannot say positively. He, like Mr. Layard, thinks that the birds must have been born on the islands in which they were found. But the hypothesis that the young birds leave New Zealand early in March, and, passing through New Caledonia, reach the Solomon Islands early in April, would fit the facts very well.

It is stated that the tuis never lose an opportunity of persecuting the long-tailed cuckoos. The tuis seem to have a natural aversion to the robbers, probably because they fear that their nests will be robbed, and that they themselves will be saddled with the burden of rearing another bird's young. Mr. W. W. Smith has seen tuis utter a wild-alarm call, boldly assail a cuckoo, and pursue it through the bush. As the cuckoo is able to offer only a feeble resistance to a number of angry tuis, it seeks safety in flight, and its superiority in this respect soon takes it out of danger, at any rate for the time being.

This species has a shrill, clear, and piercing cry, which is repeated at intervals. Its habits generally are supposed to be similar to those of the shining cuckoo.

A long, exhaustive, and interesting paper on the habits of the long-tailed cuckoo was read by Dr. R. Fulton at the meeting of the Australasian Association for the Advancement of Science at Dunedin in January, 1904. He stated that this bird generally chose an open or cup-shaped nest in which to deposit its egg, and not a domed or covered-in nest. He found that the commonest host in both islands was the native canary, but the tui, the robin, the tomtit, the white-eye, and some of the imported birds were also imposed upon. He had received abundance of evidence as to the long-tailed cuckoo's robbing propensities, but he expressed an opinion that this was a comparatively modern failing, due to the prevalence of the nests of small imported birds in the trees on the edge of the bush, where the cuckoo usually shelters. Dr. Fulton described its semi-nocturnal habits, and added: "The moment our cuckoo shows itself in the daytime it is pounced upon by all the small-fry in the way of native birds, who pursue and torment it until it reaches the safety of the long grass or thicket. Sometimes, when the cuckoo is chased by the tui, it will settle along a good-sized bough, and, turning towards its pursuer, will make a curious, defiant, crowing sound. It is noiseless in its flight, and this, no doubt, is the main reason for the success the female achieves in getting her eggs into the nests of other birds.''

ORDER PSITTACI.

Bill strong, hooked, the base covered with skin. Legs short, the feet with two toes in front and two behind.

Key to the Families.

1. Lower mandible nearly straight.   Nestoridæ
   Lower mandible much hooked.   2
2. Tail long.   Psittacidæ
   Tail short.   Stringopidæ.

Family Nestoridae.

Bill longer than deep. Tongue fringed near the tip. Shafts of the tail feathers projecting. New Zealand only.

Genus Nestor.

Hook of the bill nearly smooth below, the upper margin grooved. Wings long, rather pointed. Tail moderate, squared at the end. New Zealand and Phillip Island.

Key to the Species.

Top of the head grey.   N. meridionalis.
Top of the head green, like the back.   N. notabilis.

The Kaka.

Nestor meridionalis.

Olive-brown. Top of the head grey; abdomen and over the tail purplish red. At the back of the neck a ring of yellowish red. Inner webs of the primaries banded with red. Eye dark brown. Length of the wing, 11 in.; of the tarsus, 1.2 in. The sexes are alike. The young may be distinguished by the red feathers on the under surface of the wing being barred with brown. Egg—White; length, 1.75 in. Both Islands.

The kaka is very playful, very sociable, and very noisy. The top of its head is grey, but in this respect alone does it resemble the renowned namesake of its genus, the cunning Pylian orator, whose tongue poured forth a flood of more than honey-sweet discourse upon the strife of kings.

Living on and among trees, the kaka remains in the forest all the year round. When disturbed in its leafy home, Mr. Potts says, it hops amongst the branches with much dexterity, beak and wings assisting its awkward but rapid progress as it threads its

(*Buller.*)

Kea.   Kaka.

way amongst the leaves and sprays with unruffled plumage. "The peculiar formation of its grasping feet enables it to perform wonderful feats of agile climbing. A sharp note or two marks its uneasiness when a viligant eye watches what takes place below. When really alarmed, after a few hurried movements, it flies a short distance away, at first usually gliding downwards rather than flying straight, threading the leafy maze of the close-growing trees with perfect ease and grace." At this time it warns its fellows of impending danger by uttering loud cries of "kaka, kaka!" which are often repeated. It is from this cry that it has derived the name given it by the Maoris, which has been adopted as a popular one by Europeans.

In its habits, it is gregarious; and it is sociable even in distress, large numbers gathering round a fallen member of the company, with apparent demonstrations of commiseration. This habit led to scores being slain where otherwise only a single bird might have been victimised. The Maoris took full advantage of the habit in obtaining large supplies. They also attracted kakas to their death by imitating their cry, which can be done without much difficulty. In the "taki" system of catching kakas, a long pole was stuck in the ground, in a slanting position. A man then hid himself in a hut, made of the leaves of the fern-tree, at the foot of the pole, and either used a tame kaka as a decoy or imitated the kaka's cry. As the bird alighted on the pole and descended, turning from side to side, the man put forth his hand and caught it unawares. When a decoy was used, the end of the pole was placed inside the hut, and the unwary bird walked right in.

In dull and moist weather, the kaka is more noisy than usual. Its voice is heard at the earliest dawn, and even at night it is not silent. "Often in the bright sunshine," to quote Mr. Potts's words, "scores may be observed with loud screams and clatter, flying and circling about, and, high above the outskirts of the bush, apparently bent on enjoying some short excursion. Now and then an individual, more hilarious than his fellows, after a somewhat slow and laboured ascent, will suddenly dart downwards perpendicularly, with almost closed wings. This feat is doubtless performed to an appreciative and admiring circle, if one may judge from the clamour of the company."

It has been found that kakas, when migrating from one part of the country to another, fly at a considerable height, uttering at intervals a brief note that sounds something like "t-chrut, t-chrut," then, perhaps a whistling call of "tweetie, tweetie." They do not travel in large flocks; but two or three, and sometimes six or seven, are often found in company. Their steady, slow, and somewhat laboured flight when journeying, it is stated, cannot be mistaken for that of any other native bird on the wing. When serious business is on hand, they have a methodical, painstaking style, which is in contrast to their gay, rattling, off-hand soaring and gliding about the bush when on pleasure bent.

When a pair of kakas have mated, they are seen constantly together; if one moves from a tree, its partner follows quickly. In regard to its home, it is a troglodyte, a word which was applied by the ancients to some tribes that lived up the Nile, and which literally means dweller in a cave. The kaka, when house-hunting, generally selects a tree with a heart that is quite decayed. There must also be a convenient hole leading from the outside to the bottom of the hollow. The interior may require some preparation, and the entrance may need smoothing and enlarging. As the pair are very solicitous for the comfort and safety of the little ones expected, they are often fastidious in making these preparations. It often happens that, after the home has been prepared, and is ready to be occupied, it is deserted for a better site. Four white eggs are laid on the decayed wood, and are there hatched. Attachment to their young is a strong point with kakas. Mr. Potts adds that he has found the old bird dead at the entrance of its nesting hole, after a bush fire, in which it had perished sooner than desert its helpless offspring, though escape was easy.

In the summer, the kakas are occupied with the cares of providing for their young and protecting them. After the young are able to get along by themselves, as autumn advances, the old birds become very fat. "It is in winter time that they appear to the greatest disadvantage, especially in the south, where the snow covers the land. With ruffled feathers, they sit moping and nearly silent, a picture of dull melancholy. Towards the close of winter they have been known to devour with avidity the hard seed of the kowhai. At this time gardens and shrubberies are visited, and the blossoms of almond trees and flowering shrubs are eagerly ransacked. But, as winter passes away with its coarse fare, returning spring restores the kaka's sprightliness, and it fares daintily."

The kaka sips the honey from the flowers, and this, as well as insects, constitutes part of its diet. In September it has been seen in Canterbury, poised on the slender boughs of a tall panax, luxuriating on the viscid nectar of the blossoms. It has been accused of injuring trees by stripping the bark in its untiring search after insects, in order to satisfy its almost insatiable appetite. Mr. Potts, enthusiastic champion of the birds, has come to the defence of the kaka in this matter. He points out, in the first place, that, being a honey-eater, it may cause the fertilisation of the blossoms of trees and assist in their propagation. As to the destruction of trees, his observation has led him to conclude that it is only the apparently vigorous, but really unsound, trees that the kaka damages. Those trees are already doomed by the countless multitudes of insects, and the kaka merely hastens the end.

About the year 1856, kakas invaded Otago in such large numbers as to become almost a plague. It is stated that not only in the bush, but in the open, on stacks, fences, or the ridges of houses, they could be seen perched in rows as close as they could sit. They were seen sitting on a post-and-rail fence in the Tokomairiro Plains, in Otago, so close together that new arrivals had to fight for perching room, and, if a person shot along the line of fence, he could knock over half-a-dozen at one shot. On the occasion of that migration, they caused a great deal of damage, especially to stacks and thatched houses. Settlers, thinking that the pest would increase year by year, seriously discussed what means should be taken to deal with the birds. The following year, however, hardly any kakas were seen in the district, and the visitation has never been repeated.

The kaka's tongue is thick, fining down towards the point, not unlike a finger. The superior side is flattish, the under side rounded and furnished with a row of short stiff papillæ, black in colour. This brush-like apparatus, Mr. Potts adds, can scarcely be said to form the termination of the tongue; it really occupies a similar position on the tongue to that which the margin of the nail occupies on the human finger. On the inside of the lower mandible there may be observed, just within the deeply channelled lip, a row of minute yellowish dots, very slightly raised above the surface of the mandible. At the sides these specks give way to very faintly marked furrows, probably to clear the papillæ by the pressure of the tongue against the lower mandible. The upper and lower mandibles are connected by a thin tough skin, which allows the beak to open widely, and gives great freedom to the movements of the lower mandible. About the middle of this skin, in a line with the gape, there is a hollow sac, or pouch, containing a wax-like substance.

Mr. W. T. L. Travers, in his observations of the kaka, was struck with its inquisitiveness and absolute fearlessness. On his cattle station at Lake Guyon, in the Nelson province, he has seen several kakas gravely take post upon some tree close to him, eyeing him with the utmost apparent curiosity, and chattering to themselves, as if discussing the character and intentions of the intruder. After a lapse of a few minutes they have darted away, uttering loud cries, as if proclaiming to the rest of the forest the presence of a stranger, to be avoided or not, as the case might be. During the winter, the birds often unhesitatingly entered the house for food, making themselves thoroughly at home, and even roosting on the cross-beams in the kitchen on specially inclement nights. Two of them in particular soon learnt how to open the door of the dairy, which they were fond of getting into, in order to regale themselves on cream and butter, both of which they appeared to appreciate excessively.

Mr. Travers had several of these birds billing on the eaves of the house in the evening, waiting to be fed, and coming readily to receive from the hand pieces of bread thickly spread with butter, and strewn with sugar. But they rarely accepted any of the bread, dropping it as soon as they had cleared off the butter and sugar. If one bird happened to have finished his portion before the others, he unhesitatingly helped himself to a share of some neighbour's goods, which was always yielded without the slightest demur.

They were fond of raw sheep, and he has seen them hovering in front of a sheep's pluck hung on a tree, just as a humming bird hovers in front of a flower, eating fragments which they had torn off, giving the preference to the lungs. When anxious to get into the house, they took up positions on the window-sills and beat at the window with their beaks until they were admitted. They were very mischievous, and nearly always cut all the buttons from any article of clothing that might happen to be left within their reach. In some instances, their familiarity degenerated into such gross impudence that Mr. Travers's manager was obliged to kill them in order to put an end to their mischief.

The Kea.

Nestor notabilis.

Dull olive green, each feather edged with black. Over the tail and below the wings, red. Outer webs of the primaries, blue; their inner webs banded with yellow. Tail, green with a black bar near the tip. Eye black. Length of the wing, 12.5 inches; of the tarsus, 1.5 inch. The sexes are alike. The young have the dark edgings to the feathers broader. Egg—White; length, 1.75 inch. South Island.

The kea is tame by nature, and mischievous by inclination. It is stated that the young birds are so tame that, if a person meets with a flock of them and keeps still, they will walk up to him and pull his clothes. When the young are taken, they are easily tamed, and they have been taught to imitate the human voice. They strongly resent attempts at captivity, and some rather marvellous escapes from imprisonment have been related. A captured kea was once placed on the floor under an inverted bucket. The places for the handles would not allow the rim of the bucket to touch the floor. Taking advantage of this, the prisoner wedged his long beak through the aperture, and, using its head as a lever, raised the bucket and escaped. Another prisoner was known literally to eat its way out of a wooden cage. It brought its powerful beak to bear on the structure, and performed some remarkable feats of carpentry, until escape was effected. It is recorded that two keas, which were tamed by their captor, were allowed to wander at large, but they returned to the house with marked regularity, and then went off on their rambles again, scrambling and clamoring amongst the trees and out-buildings. Any kind of food seemed to meet with their approval; but a piece of raw meat was what they liked best.

It is well known that, to settlers in many parts of the outlying districts of Otago, Southland, and Canterbury, the kea has become a source of great anxiety, because of the evil habit it has

acquired of attacking sheep and eating the kidney-fat. Alighting
on the back of a sheep, the bird digs its beak through the skin
and flesh until it secures the morsel, and makes its cruel repast.
Very startling figures, showing the destruction caused by these
birds, have been published. Although some of the statements
seem to be exaggerated, there is no doubt that the settlers' flocks
have suffered heavily. It is recorded in the reports of the Lands
Department that, on a station at Lake Wanaka, 200 sheep were
killed in one night; while on runs far back and at high altitudes,
20 per cent. of the flocks have succumbed to the birds' ravages.
The local governing bodies and the Government still pay for
keas' heads. Some settlers have offered as much as 10s. a head
This system of outlawry has considerably thinned the numbers
of the keas, but they are reported in 1922 to be as plentiful as
ever. Some settlers slay their hundreds, others their thousands.

The kea's peculiar method of covering the ground has been
noted. It goes along with rather a hopping jump than a walk,
which gives its gait an odd appearance. For a nesting place, it
seeks the shelter of almost inaccessible rocks, or burrows in a hole
in the steep facings. It breeds very early. Its egg is larger than
that of the kaka, and rougher, and the surface is granulated with
pits dotted over, while there are a very few slight chalky
incrustations towards the smaller end. The shell is very stout
and thick. The egg is ovoid in form.

One of Mr. Potts's best chapters in *Out in the Open* deals with
this bird and its quaint habits. He says: "It breeds in the deep
crevices and fissures which cleave and seam the sheer facings
of almost perpendicular cliffs that in places bound, as with
massive ramparts, the higher mountain spurs. Sometimes, but
rarely, the agile musterer, clambering amongst these rocky fastnesses, has found the entrance to the 'run' used by the breeding
pair, and has peered with curious glance, tracing the worn track
till its course has been lost in the dimness of the obscure recesses
beyond the climber's reach. In these retreats the home or
nesting place generally remains inviolate, as its natural defences
of intervening rocks defy the efforts of human hands, unless

K

aided by the use of heavy iron implements that no mountaineer would be likely to employ.''

Another extract may be made from Mr. Potts's works to show the nature of the country usually inhabited by the kea. The writer says: ''Where I have most observed it has been above the gorge of the Rangitata, one of the great snow rivers. This stream, which derives its source from the glaciers embedded in the gloomy and secluded fastnesses of the Southern Alps, is periodically swollen by the melting of the snows and by the heavy north-west rain that falls during the spring and autumn months. Fed by numerous creeks, and tributaries from every converging valley, its volume increases, and it rushes noisily and impetuously over its rough boulder-bed till the junction of the Havelock, the Lawrence and the Clyde swells its waters into a large river. The lofty, rugged mountains which imprison it present almost every conceivable variety of outline. There are jagged peaks crowned with snow; and countless moraines show where the avalanche and the snow-slip have thundered down into the valley below. The river is bordered here and there by grassy flats and hanging woods of timber trees, in which the brown-tinted totara, the silvery parsley-pine with its purplish points, the small-leaved kowhai, and the soft, bright-foliaged ribbonwood contrast well with the dusky hue of the dark-leaved fagus. Far above, dwarf vegetation in all the wonderful variety of alpine shrubs and flowers struggles up the steepest slopes, adorning the frowning precipice and foaming cascade, lending its aid in forming scenes of picturesque and romantic grandeur, in which rich and varying tints of perennial verdure gratify the eye of the spectator with their beauty. This is the home of the kea, amongst holes and fissures in almost inaccessible rocks, in a region often shrouded with dense mists or driving sleet, where, at times, the north-west wind rages with terrific violence.''

Family Psittacidae.

Genus Cyanorhamphus.

The Parrakeets.

Bill deeper than long, not notched, bicolor, the hook of the upper mandible with a file-like surface below. Tongue simple. Second and third quills the longest, the outer webs of the second to the fourth strongly sinuated. Tail long and graduated. Colour green. New Zealand, New Caledonia, Loyalty Islands, and Society Islands.

Several species of parrakeets belong to New Zealand, and to the islands near the mainland. They are exceedingly beautiful birds. Some of them have been taught to imitate the human voice fairly well. In the early days, they swarmed in many parts to such an extent that they became a nuisance to the settlers. Writing, in regard to Otago, in 1877, Mr. R. Gillies says: "To say that they are continually to be seen in flocks of hundreds gives a very faint idea of their extreme prevalence. Settlers whose cultivations were in the bush had always the greatest difficulty in saving crops of wheat. These lively, roguish little birds defied all scarecrows, and even shooting was found to be an endless and expensive job, for, though a few might be killed at a blow, the flock just rose and settled down again immediately a few yards off. I have known patches of wheat rendered utterly valueless by the parrakeets. So rare and scarce have they become now that country settlers near the bush have quite a warm side to the little green parrakeets, and often make household pets of them."

In 1885, it was reported that there was a plague of parrakeets in the Mackenzie Country, and a run-holder near Burke's Pass shot 100 in two days. This slaughter, however, seemed to have had no effect, and the unfortunate man's fruit crop was utterly destroyed. In 1888, they invaded the towns and villages on the West Coast, and spread themselves over the settled districts of Canterbury As the settlers had to take steps to protect themselves, the birds perished in thousands, but not before large quantities of fruit had been consumed. Later on, in reduced numbers, they returned to the bush, and to their natural haunts.

Mr. W. W. Smith, in a letter on the subject published in the *Lyttelton Times* at the time, attributes the disturbance mainly to a partial or total failure of ordinary food supplies. "This contention," he says, "is fully borne out by the miserably lean and generally starving condition of the birds when they first arrive in the district, and by the poor state of the plumage, and the presence of numerous parasites on their bodies. In New Zealand, in seasons of failure of the indigenous berry-bearing trees and shrubs, the parrakeets are not the only birds that suffer, as the failure affects, with a few exceptions, all other bush birds. A season of scanty blossoms produces a scarcity of the insects depending on them for support, and this affects insectivorous birds. The same effect is felt by the honey-suckers, such as the kaka, the tui, the bell-bird, and others, which, during the spring and summer months, depend for much of their nutriment on the melliferous flowers of the bush." Mr. Smith also states that simultaneously with a previous invasion of parrakeets, the towns and villages on the West Coast were invaded by an army of rats. These animals, which subsist during several months of the year on the fallen berries of many forest trees, were driven from their haunts in the bush by precisely the same cause as that which affected the parrakeets.

Key to the Species.

1. Forehead, green.	C. unicolor.
Forehead, orange.	C. malherbei.
Foreheal, red.	2
2. Tail, partly blue.	C. cyanurus.
Tail, entirely green.	3
3. Crown, red.	4
Crown, yellow.	5
4. Upper surface, yellow-green.	C. erythrotis.
Upper surface, grass-green.	C. novæ-zealandiæ.
5. Crimson forehead reaching the eye.	C. auriceps.
Crimson band of forehead not reaching the eye.	C. forbesi.

The Antipodes Island Parrakeet.

Cyanorhamphus unicolor.

Green, the base of the outer primaries bluish. Eye crimson. Length of the wing, 6 in.; of the tarsus, 1 in. The female is slightly smaller. Antipodes Islands. The wing is shorter in proportion to the length of the tarsus, than in the other species. The crest of the sternum has about the same proportion of length to breast as in *C. erythrotis.* It feeds largely on the seeds of the *Acaena sanguisorbae.*

Members of this species, as well as of *C. auriceps* (p. 151) were placed on the Kapiti Island sanctuary by Dr. L. Cockayne in December, 1907.

The Kermadec Island Parrakeet.

Cyanorhamphus cyanurus.

Green, the top of the head crimson. Tail blue, the two central feathers tinged with green. Length of the wing, 6.6 in.; of the tarsus, 0.82 in. Raoul Island, Kermadec Group.

The Red-fronted Parrakeet.—KAKARIKI.

Cyanorhamphus novae-zealandiae.

Bright green with blue on the wings. Forehead and top of the head crimson. Variable in size. Eye crimson. Length of the wing 5.2 to 4.6 in.; of the tarsus, 0.75 to 0.56 inch. The sexes and the young are alike in plumage, but the female is smaller than the male. Egg—White; length, 1.1 inch. Both Islands, and Auckland Islands. In the Auckland birds the depth of the crest of the sternum is less in proportion to its length than in birds from New Zealand.

Many years ago the red-fronted species was numerous in the beech forests in the Malvern district and in the smaller woods along the banks of the alpine riverbeds of Canterbury. It became very scarce for a number of years, and then increased again. Breeding early, it commences nesting almost in the winter. Its nesting-places are found at various heights, from a few feet to at least 60 feet from the ground. A hole in a shallow tree or a decayed branch is sometimes selected, and occasionally it is placed between the hole of the tree and a piece of loosened bark. The eggs are deposited on the decayed wood at the bottom of the hole, or a slight nest is formed. One nest obtained was made entirely of feathers, moss, and the downy scales of tree-ferns. These materials were arranged into very slight fabric of a cup-like form, which just fitted the hollow selected for the breeding place. The hen lays five, six, or more eggs, which are broadly oval, and sometimes nearly spherical. At nesting time the birds often indulge in a low murmuring note to each other.

Yellow-fronted Parrakeet.

Red-fronted Parrakeet.   *(Buller.)*

The Yellowish Parrakeet.

Cyanorhamphus erythrotis.

Yellowish green with blue on the wings. Forehead and top of the head crimson. Length of the wing, 5.6in.; of the tarsus, 0.82in. Macquarie Island and Antipodes Island. There is no reduction in the length of the wing; but the crest of the sternum is more reduced than in the Auckland Island parrakeets; nevertheless, it flies well.

The Yellow-fronted Parrakeet.—KAKARIKI. *

Cyanorhamphus auriceps.

Bright green with blue on the wings. Forehead, crimson; top of the head yellowish orange. Length of the wing, 4.5 in.; of the tarsus, 0.75 in. Egg—White; length, 0.9 in. Both Islands and the Auckland Islands.

The yellow-fronted parrakeet is a specially beautiful bird, and it has a merry note. It rears its young in a hollow tree or branch, and the nest is sometimes placed between the wood and the dissevered bark of a decaying tree, but more often at the bottom of a deep hole. The eggs are somewhat oval in shape.

The Chatham Island Parrakeet.

Cyanorhamphus forbesi.

Similar to C. auriceps, but larger, and with the crimson band in front of the yellow crown much narrower. The crimson band does not reach the eye. Chatham Islands.

The Orange-fronted Parrakeet.

Cyanorhamphus malherbei.

Bright green with blue on the wings. Forehead, orange; the crown, pale yellow. Length of the wing, 4.1 in.; of the tarsus, 0.66 in. Egg— Not known. South Island and Little Barrier Island.

Family Stringopidæ.

Bill, thick, swollen on the sides, not notched, under side of the hook with a file-like surface; the lower mandible grooved. Base of the bill covered with feathers, the shafts of which are prolonged into hairs. Wings short and rounded, the fourth and fifth quills the longest. Tail short and rounded, the end of each feather pointed. Tarsi rather long and strong. New Zealand only.

Genus Stringops.

The same characters as the family.

---

*Archdeacon Williams gives the Maori name of this bird as "Powhaitere."

Kakapo.   *(Buller.)*

The Kakapo.

Stringops habroptilus.

Above, green varied with brown; below, yellowish green varied with brown and yellowish white. Eye black. Very variable in size. Length of the wing, 11 in.; of the tarsus, 1.75 in. Egg—White; length, 1.9 in. Both Islands, but very rare in the North.

Size, appearance, and habits of life combine to make the kakapo one of the most remarkable birds in the dominion. Its intelligence commands respect, and its helplessness sympathy, while its genial nature endears it to all who know it well. It repays kindness with gratitude, and is as affectionate as a dog, and as playful as a kitten.

Almost every writer who has written about the kakapo has dwelt upon these characteristics. Sir George Grey, writing at an early date, says that its behaviour towards friends is more that of a dog than a bird. Professor G. S. Sale has seen it run from the corner of a room, seize his hand with claw and beak, and tumble over and over with it, just as a kitten would. It then rushed back again, so as to be invited to make another playful attack. The professor was amused by the kakapo's humour when a dog was placed close to its cage. It first danced backwards and forwards with outstretched wings, evidently with the intention of shamming anger, and showed its glee at the success of the manœuvre by assuming absurd and grotesque attitudes. When pleased, it marched about with its head twisted round, and its beak in the air, wishing, the professor presumed, to see how things looked the wrong way up. The highest compliment it could pay to anyone was to nestle down on his hand, ruffle out its feathers, and lower its wings, fluttering them alternately, and shaking its head from side to side. When it did that, it was in a superlative state of enjoyment. He says that its principal resorts are the grass plots in the open and mossy beech woods near the mountain streams, rocky declivities, beneath large moss-covered stones, overgrown by beech-roots, and the mossy banks of large rivers.

Being mostly a night bird, the kakapo spends a great deal of the daytime in holes in the ground, sometimes under the roots of trees. Its wings are both large and strong, but their muscles are so weak that the bird cannot fly. In diet it is a gluttonous vegetarian, feeding on grass, weeds, fruit, seeds, and roots. When taken from one of its holes, it will endeavour to hide again, as soon as possible, in its dark retreat. The use of the wings is not altogether discarded, as they are brought into requisition when the bird is running, and are also used on trees as a means of balancing. In making a descent, the bird sometimes half jumps and half flies, but on some occasions it drops to the ground like a stone. Ascents of trees are made by a climbing habit, beak, claws and wings being used. When feeding, and when pleased with its food, it emits a curious grunt. Its general cry has been described as not altogether a shrill scream, but a muffled screech, more like a mingled grunt and screech.

The plumage is remarkable because it resembles in appearance the moss in which the bird finds large quantities of its favourite foods, so that, when seeking protection from enemies, it has some kind of compensation for the weakness of its wings.

The literature in connection with the kakapo commences in 1852, when Dr. Lyall published a long article on the subject in the *Proceedings* of the Zoological Society, London. Since that time, many other articles have appeared in various publications. One of the best was written by Sir Julius Von Haast, and published in the *Verhandlüngen* of the Zoological and Botanical Association of Vienna, on October 10th, 1863. The writer describes his acquaintance with the bird on the west coast of the South Island.

"The black hairy feathers on each side of the beak," he says, in describing the bird's appearance, "give it a somewhat wild appearance; and the curious radiating wreaths of feathers round the eyes make it look like an owl; but the large parrot-like beak and the two reversed toes determine at once its affinities." The first time he saw the kakapo by daylight was on an afternoon. The bird was sitting on a stump in an open part of the forest, not far from the Haast River. On his approach, it quickly disappeared, but was soon caught by his dog. On another occasion, he saw a large specimen in the daytime in a mountain pass. It was sitting on a fuchsia tree, ten feet from the ground, eating berries. As soon as it saw him, it threw itself off the tree as if it had been shot, and escaped under some large fragments of rock. He was greatly surprised to note that it did not open its wings or use them in any way to break its fall.

To see whether or not the kakapo would fly or flutter when pursued, he had a large specimen brought to an open place, where there was sufficient space for it to open its wings whilst running. Instead of doing that, the bird, when released, ran towards the nearest thicket, moving like a fowl, with a speed which, considering the position of its toes and its unwieldy form, surprised him.

The crops of those he examined were generally filled with minutely divided moss, in enormous quantities. They were greatly distended, and were sometimes so heavy that a single one weighed several ounces. "The mass of this little nutritious food which the bird must collect," Sir Julius says, "shows why it lives on the ground, and in barren and unproductive districts where no other species of the same family could exist. Another peculiarity, perhaps likewise resulting from its vegetable diet, is that the bird, instead of having, like others, an oily, soft kind of fat under the skin, possesses a great quantity of firm white fat. Its flesh is better and more substantial than that of any other species of parrot, and is of exquisite flavour.

"I expected to find the kakapo in well-excavated caves, with entrances which would permit the inhabitants to enter, something like the lair of a fox or a badger. But I found, that, with the exception of a single instance, the habitations consisted of clefts or fissures in rocks, holes between the roots of decayed trees, or natural openings between fragments of rocks, where my large dog entered easily, and generally returned head first, carrying the prey in his mouth, showing that he must have been able to turn round within.

"At first my dog was severely punished by the beak and claws of the kakapo; but, after a little experience, he learned how to grasp the bird through its head at once. The Maoris told me that the kakapo was a very valiant bird, and often fought successfully with their dogs; but this is scarcely credible, unless their dogs are a very weak race. My dog, though punished at times, never had a serious battle with one of them. All the habitations of the kakapo that I examined were natural caves or holes, with the exception of one, which was artificially excavated." Sir Julius's observations led him to the conclusion that the birds lived singly, but went about in pairs at night.

While in New Zealand, Mr. Reischek devoted special investigations to the manner in which the kakapos made their tracks, which have been found in many parts of the dominion, and were sometimes mistaken for those of human beings. He writes as follows: ''It was very amusing to watch these creatures, generally one at a time, coming along the track feeding, and giving a passing peck at any root or twig that might be in the way. Thus the tracks are always kept clean. In fact, they very much resemble the native tracks, except that they are rather narrower, being from eight to fourteen inches wide. The kakapos generally select the tops of spurs for the formation of their tracks. I was curious to know how the birds would manage when the ground was covered with snow. I found that they travelled on the surface of the frozen snow, and that the tracks were plainly visible. In many places the scrub, which consisted of silver pine, ake-ake, and other alpine vegetation, was so dense that the snow could not penetrate it. The kakapos take advantage of this to make their habitations under the snow-covered scrub, where it is both dry and warm.

''They leave their burrows after sunset, and return before daylight. If they cannot reach their own homes during the darkness, they will shelter in any burrow which may be unoccupied, as they travel long distances. They consume large quantities of food, which consist of grass, grass-seed, and other alpine vegetation. In July they are in splendid condition, those found having as much as two inches of fat upon them. The young birds are delicious food when roasted in the camp oven.

''In the spring when the sun begins to shed its warmth, the kakapos emerge from their burrows, and select some favourable spots in the sunshine, where they crouch down and remain the whole day. In September, I selected a suitable day for observing this peculiarity. The snow had disappeared from all the sunny places. I found three birds in different places sitting upon low silver-pine scrub. They took no notice of my approach until I had them safely in my hand, when they endeavoured to release themselves by biting and scratching. The bush kakapos, as well as the alpine ones, become very fat in the winter months.''

ORDER RAPTORES.

Bill hooked. Feet strong, three toes in front and one behind; all armed with strong, sharp claws, which are more or less retractile.

Key to the Families.

No facial disc.	Falconidæ.
With a facial disc.	Strigidæ.

Family Falconidae.

Base of the bill covered with skin. Plumage compact, wings long and pointed. The legs naked.

Key to the Genera.

Upper mandible with a sharp tooth.	Nesierax.
Upper mandible sinuated.	Circus.

Genus Nesierax.

Upper mandible with a sharp tooth, the nostrils round. Third quill the longest. Outer toe (without the claw) longer than the inner toe (without the claw) ; middle toe very long. New Zealand only.

The Quail-hawk.—KAREWAREWA or KAREAREA.

Nesierax novae-zealandiae.

Above, brownish black. Below, rufous-brown, spotted with rufous-white; chin and throat, white with dark brown streaks. Thighs, rufous, streaked with brown. When very old, the upper surface is banded with brown, and the breast is rufous, with brown streaks. Very variable in size.—(See Ibis, 1873, p. 101, and 1879, p. 460.) Length of the wing, male, 9.5 to 10.5 inches, female, 11 to 12 inches; of the tarsus, male, 2 to 2.4 inches, female, 2.3 to 2.6 inches. Eye hazel. Egg—Yellowish white, mottled all over with rich reddish brown; length, 2 inches. Both Islands and Auckland Islands.

The quail-hawk, a true falcon, with a distinctly toothed bill, stands at the head of the diurnal birds of prey. In this species, as in all falcons, there is a great difference in the sizes of the sexes, and in the colours of the young and of the adult. This has led to

much confusion and difference of opinion as to whether the
dominion possesses two species or only one. The general opinion
seems to be that there are two, namely, the quail-hawk, and a
smaller bird, the bush-hawk. Both are found in the Auckland

Quail Hawk.   *(Voy. au Pole Sud.)*

Islands, but it is stated that the larger form does not
occur in the North Island of New Zealand, though both used to
be common in the South Island. It is also stated that the quail-
hawk breeds on the plains and lower ranges of the hills, while
the bush-hawk breeds near the summits of the wooded ranges.

The quail-hawk, when in pursuit of its prey, is courageous and persevering. It has made a bad reputation for itself as a daring marauder. It will swoop down on a farmyard, strike a fowl, and cling to it until knocked over with a stick. It seems to be quite insensible to danger, darting into houses, and following its prey into inner rooms. This bold mountaineer specially delights in capturing domestic pigeons, which afford a tasty meal. Having secured its prey, it takes it to a sheltered spot, plucks it rather carefully, and devours it at once.

Mr. Potts has given a vivid account of how the hawk pursues a victim it has selected for a meal. "In one of the bays of Banks Peninsula," he says, "a range of hills, curved in outline, stands back some distance from the sea. On the steep rocky crags above, the falcons once had their eyrie. Often the locality was visited by them years after the old breeding-place had been deserted. Many a time when the quail-hawk has dashed across the peaceful bay in its swift career, I have watched its course with the utmost interest. I have noticed that, when the pigeons were alarmed at the shrill garring screams of the falcon, they swept round the bay at a great height. Soon a flock seemed fluttered. The falcon is amongst them; birds dart from the rest of the flight; presently a victim is singled out, and cut off from the remainder of the bewildered birds.

"Then ensues a contest for the upper air, pursuer and pursued striving for the advantage of being uppermost. Soon the stronger muscles and the superior tactics of the falcon obtain the mastery. The pigeon needs all its speed and skill to avoid the fierce strokes of the sharp, hooked talons; adroitly he eludes one or two strokes, then seeks for safety in an onward flight. This trial of speed often results in the escape of the pigeon. Its swift pinions make it almost a fair match for the hawk. When nearly exhausted, the pigeon avails itself of any covert to drop out of sight of its relentless enemy. In the fierce rush of wings, the falcon, with wonderful dexterity, surmounts opposing obstacles, such as buildings, against which one feels it must surely destroy itself, taking them with an easy bound. If his quarry lies perdu, he hies him off to a fresh venture. On several occasions our pigeons have been thus molested, and many a good flier has succumbed to the spoiler. A quail-hawk has remained about the neighbourhood for ten days or a fortnight, that is, till he has exhausted the patience and forbearance of some plundered poultry keeper. Probably in that time four or five pigeons have been struck down.''

Having firmly secured its spoil, the quail-hawk will not readily let it go. It is recorded that, at a cattle muster in the Upper Ashburton district on one occasion, a female bird was seen with a tui trussed in her talons. An attempt was made to make the hawk drop the smaller bird, but without success. Neither stones nor the loud report of a stockwhip had any effect, and the hawk flew a long distance without a rest, still holding its prey. The male bird soared boldly in company, keeping watch over his companion and her precious freight. The hawk has been seen to attack seagulls in Milford Sound and Preservation Inlet, and quail and wood hens often fall before the dreadful tyrant of the air. A wood hen which was picked up at its last gasp showed that the fatal stroke had been dealt on the head and neck, from which only a few feathers had been displaced.

It is thought that the hawk's pursuit after different species of birds must call forth entirely different tactics and methods. The chase after the noisy, screaming kaka, often wheeling in its laboured flight, turning upwards its strong beak and claws in attempts to ward off the impending stroke, must differ from the chase after the silent, strong-winged pigeon, and each of these must be different from the method of attacking a parrakeet.

Nicety of calculation in regard to the force of the stroke and neatness of execution are said to be displayed at their best when the hawk picks off a kingfisher while it sits on a telegraph wire. An observer has seen the wary kingfisher, discarding its accustomed shyness, save itself by dashing, with a wild shriek of terror, amongst a party walking round a garden, its baffled enemy coolly rising above the close-set shrubs and sailing off deliberately towards a good look-out station on the bare, spreading limb of a lofty tree.

The Australian magpie has been known to defend itself against a hawk's attack by throwing itself on its back on the ground, and striking out with beak and claw, and shrieking dismally. The quail-hawk has swooped down at a lark, struck fiercely at a large harrier, and ended his sudden foray, all in a few minutes, with the death of a good-sized chicken. When it secures a fine catch, it gorges itself to repletion. Though it plucks birds before devouring them, it rends small prey, such as rats and mice, into gobbets, and these are swallowed fur and all.

"The breeding-place of this remarkable bird is usually on a ledge of rock commanding a prospect over some extent of country. The site is well chosen, and it gives the bird an excellent outlook, and affords a very advantageous position for a creature of its habits and inclinations. Bold rocks, somewhat sheltered by projecting or overhanging masses, appear to be favourite sites in which to rear the young. There are generally three eggs, which are deposited on any decayed vegetable matter that wind or rain may have collected on the rocky ledge. The efforts of this bird in regard to nest-building are of a very feeble description. Some eggs found were of a reddish brown colour, mottled with darker shades of brown; sometimes the ground colour is pale reddish white, less suffused with the darker colour at the smaller end, broadly oval in shape; they measured two inches in length, and had a diameter of an inch and a half. Other eggs were of a yellowish colour in place of reddish brown, while others again were of dull chocolate shades, sprinkled with fine dots and blotches of rich brown, most abundant at the larger end. It is believed that the young birds either leave the protection of the parents, or are driven off, in the beginning of autumn, when they are well grown. The breeding season extends through the months of October, November, and December."

Bush-hawk.—KAREWAREWA or KAREAREA.
Nesierax australis.

The bush-hawk, or sparrow-hawk, lives in the mountains, where the forest is low and dense. Rats, mice, lizards, poultry, ducks, and young turkeys, as well as many birds of the forest, are victims of its rapacious appetite. Though smaller than the quail-hawk, it is swifter and more savage and resolute.

L

A pair of bush-hawks have been known to assail human beings for two hours while in the neighbourhood of their young, with savage and threatening tones. Up the Waio River, at breeding time, bush-hawks have often chased cattle dogs to the shelter of the stockman's horse.

"One day," says Sir Julius Von Haast, in his *Journal of Exploration in the Nelson Province*, "while I was walking along near the margin of the forest in Camp Valley, my hat was suddenly knocked off my head, and at the same time I heard a shrill cry. On looking up I found it was one of those courageous little sparrow-hawks which had attacked me, and which, after sitting for a moment or two on a bough, again pounced on me; and, although I had a long compass-stick in my hand, with which I tried to knock it down, it repeated its attacks several times."

The same writer gives another instance of the courage displayed by these birds. A large white heron, standing in the water, was attacked by three hawks at once, and they made frequent and well-concerted charges upon him from different quarters.—"It was admirable to behold the white heron, with his head laid back, darting his pointed beak at his foes with the swiftness of an arrow, while they, with the utmost agility, avoided the spear of their strong adversary, whom at last they were fain to leave unmolested."

Another day, in the same neighbourhood, a cormorant, passing near a tree on which two sparrow-hawks were sitting, was pounced upon by them, and put to hasty flight, with a shrill cry of terror, followed closely by the small, but fierce foes; and all three were soon out of sight.

The nests of these little falcons are placed sometimes on the top of an old dead tree stem, sometimes in a hanging mass of climbers, and sometimes on the ground.

Genus Circus.

Upper mandible sinuated, the nostrils oval. Third and fourth quills nearly equal and longest. Legs slender, the hinder aspect of the tarsus reticulate. Spread over the whole world.

The Harrier.—KAHU.

Circus gouldi.

Above, brown varied with rufous; over the tail, white with a rufous bar near the tip of each feather. Tail, silver grey with brown bars. Under parts, rufous white with reddish brown stripes on the breast; the thighs, with rufous streaks and spots. Eye yellow. Length of the wing, 16.5 inches; of the tarsus, 4 inches. Young, dark brown above, varied with white at the back of the neck; below, reddish brown; the thighs, rufous. Eye hazel. Egg—white; length, 1.9 inch. Both Islands, Chatham Islands, New Caledonia, Fiji, and East Australia.

In open low country, in both Islands, the harrier is still common. It may often be seen flying with lazy pinions near the ground, or soaring into the air. It is stated that some years ago it was one of the commonest of the larger birds met with ou the plains; but the efforts of the acclimatisation societies have altered the position considerably.

This big hawk possesses characteristics quite different from those of the other two members of the family. It is not so reckless as the quail-hawk, nor so swift in its movements as the bush-hawk. "It soars aloft noiselessly," says Mr. Potts; "it seldom appears to be hurried, but floats calmly in ascending circles, with its wings so apparently motionless that it might be saluted as the albatross of the plains. Its course, unlike that of the falcon, is generally circuitous."

In its mode of pouncing on its prey, he adds, it discloses peculiar craftiness. It will pretend to pass over its victim, then suddenly turn, wheel, and rush noiselessly to the ground. It plucks its spoil carefully, and picks the bones clean. It has often been known to capture large birds, such as turkeys, and it has even attacked a weak lamb. Offal and garbage are its principal foods. One of its chief delights is to ransack a duck's nest, robbing and eating the eggs. Lizards, cicadas, grasshoppers, crickets, and other insects, as well as rabbits, rats, and mice, are also on its bill of fare.

The harrier is not courageous. It has been beaten by the falcon, the oyster-catcher, and the gull, and has been chased

The Harrier.

(Buller.)

away by a pair of Australian magpies. "When the great harrier ventures near the nesting-place of the little bush-hawks, it is furiously assailed. The bush-hawks utter a kind of neighing scream, and both male and female generally unite to beat off the intruder. Circling above their foe, they swoop upon him, and the harrier, hard pressed, turns completely over on to his back, stretching out in defence his terrible talons.'' Mr. T. W. Kirk once saw a harrier attacked and badly beaten by a flock of sparrows, at a spot between Featherston and Martinborough. He heard an unusual noise, as if all the small birds in the country had joined together in one great quarrel. Looking up, he saw a harrier, which was being buffeted by sparrows. They were there in hundreds, and they dashed at the harrier in scores, and from all points at once. The unfortunate hawk was quite powerless. He seemed to have no heart left, as he did not attempt to retaliate, and his defence was very feeble. At last, on approaching some scrub, he made a rush indicative of a forlorn hope, gained the shelter, and remained there. The sparrows crowded round the bush. They maintained a constant chattering, as if congratulating themselves on victory, and challenging the enemy to come out and fight. He, however, cowered under cover until the sparrows, tired of waiting, flew away

Harriers generally select a breeding-place in a low-lying situation, among swamps. Sometimes, however, nests have been found on a considerable elevation in a deep gully in the hills, where a batch of toi-toi grass and a few flax bushes offer a reasonable shelter for the progeny. When the nest is approached, the harrier does not make a resolute defence, as is done by the courageous bush-hawks; but, when incubating, it utters a shrill scream of alarm, and darts off.

Family Strigidae.

Eyes directed forwards and encircled by a facial disc. Nostrils hidden by bristles. Outer toe reversible. Plumage soft. Universally distributed.

Key to the Genera.

Tarsus twice the length of the middle toe.	Sceloglaux.
Tarsus not twice the length of the middle toe.	Ninox.

Genus Sceloglaux.

First primary less than the eighth; third to the fifth nearly equal and longest. Tarsi feathered, twice the length of the middle toe and claw. New Zealand only.

The Laughing Owl.—WHEKAU.

Sceloglaux albifacies.

Brown, spotted with fulvous on the breast, and streaked with the same colour on the back. Tail barred with fulvous. Feathers on the legs pale rufous white. Sometimes the greater part of the facial disc is white. Eye dark reddish brown. Length of the wing, 11 inches; of the tarsus, 2.65 to 3 inches. Egg—White; length, 1.95 inch. South Island.

The laughing owl lives in crevices of the rocks, and formerly was not uncommon in the South Island, though it is now very rare. The female is rather smaller than the male.

The peculiar cry which has given this owl its name, and which is like an uncontrollable outburst of laughter, is heard only when the birds are on the wing, and generally on dark and drizzly nights, or immediately preceding rain. It is stated that the call of the adults, in waking up in the evening, is very similar to the call of two men cooeeing to each other from a distance. The cry of the male is loud and hoarse, and that of the female shriller and less prolonged.

In former times, the principal food of the laughing owl, it is thought, was the native rat. At present, it seems, these birds live on rats, mice, and lizards, and the large species of coleoptera common among the debris beneath the rocks where they live.

It is stated that the species is rapidly becoming extinct, owing to an insufficient supply of proper food, the coleoptera being totally inadequate to support this large bird. In captivity it soon becomes tame, and superior food produces a marked improvement, the bird, in a few weeks, becoming fatter, stronger, and bigger. In both the wild and the captive state, breeding begins in September and October, and the female sits on the eggs for twenty-five days.

Mr. W. W. Smith has recorded the very interesting manner in which the male bird caters for its mate. He kept a pair in a large packing case, with a dark recess in one corner, and when the female was hatching her eggs, the male regularly carried every morsel of food into the dark recess, and fed the female, which

(*Rowley's Ornith. Miscel.*)
Laughing Owls.

was sitting on the eggs. Sometimes the male relieved the female in the work of hatching. Throughout the breeding season the birds are noiseless, rarely uttering a sound, except whenever the male is carrying food to the nest, when he utters a hoarse call, the female rising from the eggs and responding with a peevish twitter when receiving the food. When the young are hatched, they are fed by the parent birds on large blackish worms, procured from the edges of swamps.

The laughing owl is found in the bleakest tracts of country, and, unlike the morepork, described further on, it does not display a special liking for the bush. In its habits it is strictly nocturnal. Sometimes, in pursuit of its prey, it encounters the most rigorous weather of the alpine regions.

Genus Ninox.

Like the last, but with a shorter tarsus. Southern Asia, to Australia and New Zealand.

The Morepork.—RURU or KOUKOU.

Ninox novae-zealandiae.

Above, brown, spotted with fulvous; below, rufous, streaked with brown, on the abdomen. Feathers on the legs, rufous. Eye bright yellow. Length of the wing, 8 in.; of the tarsus, 1.5 in. Egg—White; length, 1.5 in. Both Islands.

Of the two owls belonging to New Zealand, the best known is the morepork. It may still be seen in the forests, and sometimes in the open, though members of the species are not nearly as numerous as they used to be.

For an owl, it is a small bird. Its popular name has been given on account of the peculiar cry it makes, resembling the words "more pork, more pork." In the light of day it spends most of its time in the gloomiest forest shades it can find. At night it comes out to hunt for prey in the shape of rats, mice, moths, insects, and small birds. When disturbed in the daytime it is dazed, and flies off to some shady spot, in an awkward and irregular manner. While at night it is a dreaded enemy of other denizens of the forest, in the daytime it is at their mercy.

The Rev. W. Colenso relates an amusing incident noted by him many years ago when the New Zealand woods teemed with bird

life very different from that which is found in them now. A
little New Zealand owl was nestling close under the fronds of a
fern-tree. As soon as his retreat was discovered by some small
birds, the battle, or rather, the mobbing, began. The incessant
noise the little fellows made brought up their friends from all
quarters, so that the observer was astonished to see the cloud of
birds gather so quickly. They were so filled with rage, and so
intent on insulting their enemy, that they were apparently utterly
fearless and regardless of the presence of a human being. But
while they would often fly up close to him, they never laid hold
of him, or touched him with their beaks, and not a feather flew.

Still, the owl did not like it, and tried hard to get at them
without removing from his perch, by thrusting forth his head and
fiercely snapping his beak. While there was noticeable a
difference in the dilation of the pupil of his eyes, which sometimes
glared on the disturbers of his sleep and peace, it is doubtful if
he clearly saw them, although he must have heard them well
enough. Occasionally, the owl, when persecuted in this manner,
flew away to some other neighbouring tree or bush, but, in doing
that, he generally made a woeful mistake, sometimes coming
abruptly against a branch, or between the close-growing canes of
supplejacks, and sometimes lighting in a less secure place, where
the enemy could surround him. Another flight would take place,
perhaps back to his old quarters; but it always seemed as if there
could be no rest, no peace, for him while daylight lasted. When
night came on, however, the tables, no doubt, were turned upon
his persecutors with heavy interest. The words of the old song
apply to him as well as to the owl of the Mother Country:

> Not a bird of the forest e'er mates with him,
> All mock him outright by day;
> But at night, when the woods grow still and dim,
> The boldest will shrink away.

There were no mice in the country before Europeans came, so
that the owl had to depend almost solely on small birds for its
animal food, though it also fed upon grasshoppers, wetas, and
other large insects.

In the Maori fable of the *Battle of the Birds,* in which the
sea-birds fought the land birds for their territory, the morepork,
it is set forth, could not take part, as the battle was fought in the
daytime, when owls cannot see. But, at the close of the day,
when the sea-birds had been driven off and defeated, the owl was
awarded the honourable position of herald, and added to the fears
of the enemy by joining in the pursuit with its insulting discord-
ant note of ironical derision:—"Toa koe! toa koe!" ("Thou
are brave! thou art victor!")

When Heke and Kawiti were making an attack on the
Europeans in the Bay of Islands, the native parties, in taking up
their positions before daybreak, communicated their movements
to one another by imitating the cry of the morepork, which the
sentries were accustomed to hear, and of which they therefore
took no notice.

In one of his long journeys through the Wairarapa district,
Mr. Colenso was benighted, and pitched his tent under a tree at
the edge of a thicket. His companions soon fell asleep, but he
remained sitting up, reading and enjoying the stillness of the
night. Presently he heard a strange noise, or, rather, a succession
of peculiar and unusual noises, such as he had never heard before.
They were repeated in different keys and in semi-discordant
tones, mingled with shrill hissing, and seemed to come from some
creatures over his head. After this had been going on for some
time, he unlaced the door of his tent, and went out. He saw two
owls on a bare extended horizontal branch of the tree, only a
few feet above him. They were a pair, apparently carrying on
their courtship in the most grotesque manner imaginable. With
a movement that appeared to be half sedate and half humorous,
the male advanced from his end of the branch, with his head-
feathers trimmed and set up cap-a-pie, and his wings hanging
down, making a jarring noise, as if he were a turkey-cock, and,
at the same time, uttering many strange sounds, high and low,
short and long. With stately and measured steps, the female
retreated to the furthest end of the branch, turned round,
bridled herself up, and hissed in a scornful manner. When the
male retired, the female would come forward very slowly, and

with much coquettishness, to her former position, and the solemn
fun would be gone through again. "Words fail me to describe
the sounds and grimaces," says Mr. Colenso; "the usual gravity
of the bird seemed to have been burlesqued, and I laughed
heartily at the serio-comic performance, after watching it for
about half-an-hour, when, as it was still being carried on without
alteration, I returned to my tent."

The morepork generally builds its nest in hollow trees, but
some breeding-places have been found in hollow rocks in the wood.
Mr. Potts relates an adventure he had with one of these birds.
It was early morning in the summer time, and the owl was sitting
on a gate. Anxious to watch and study its motions, Mr. Potts
sat down close by it. But it soon made a sudden swoop at the
intruder, and repeated the manœuvre several times, most perseveringly, and with great gravity and deliberation. After each
attack, the bird resumed its perch on the gate, and only once did
it make a blow felt. The observer rose and walked up a dark fern
gully some distance away, and the owl followed and attacked
him again.

ORDER COLUMBIFORMES.

Bill with a hard tip and a soft swollen base. Feet with three
toes in front and one behind, all at the same level.

Family Treronidae.

Tarsus generally shorter than the middle toe. Feathered for
more than half its length. The soles of the feet broad.

Genus Hemiphaga.

Bill moderate. Wings pointed, the third and fourth quills
nearly equal and longest. Tail long, composed of twelve feathers.
New Zealand and North Island.

Key to the Species.

Outer wing coverts green.	H. novæ-zealandiæ.
Outer wing coverts grey.	H. chathamensis.

Wood Pigeon.   *(Buller.)*

The Wood Pigeon.—Kuku or Kereru.

Hemiphaga novae-zealandiae.

Above, coppery purple; head, neck, and breast coppery green.
Abdomen white. Tail greenish black. Feet pink. Eye crimson. Length
of the wing, 10 inches; of the tarsus, 1.25 inch. The sexes are alike.
Egg—White: length, 1.9 inch. Both Islands.

The beautiful plumage of the native wood pigeon gives it rank among the most handsome birds belonging to this dominion. Unfortunately for itself, its flesh is very palatable, and pigeon pie is a dainty dish that is much relished.

Pigeons formerly frequented the bush in almost all parts of New Zealand. Though their numbers have been greatly reduced, they are still fairly plentiful in places. Mr. W. W. Smith reports that they were very common in the Lake Brunner district when he visited the locality, about 1890. He noted that, in fine weather, large flights changed their quarters daily, flying from shore to

Nest of Wood Pigeon.

shore, or from one part of the bush to another, to visit some favourite or seasonable berry-bearing trees. A plentiful season of miro berries is invariably followed by a season of fat pigeons.

The pigeon's nest is remarkably well constructed. It has been compared in shape to the hollow of the human hand. The materials of the slight fabric, which appear at first to be rudely and carelessly placed together, are so nicely adjusted as to bear with perfect safety the weight of the heavy builders. In the slight depression of the platform, the egg, or young, lies undisturbed by the swaying caused by passing winds. There is a long breeding season. The reason for this, it is thought, is the bird's migratory habits, which lead it to feeding grounds where its food is to be found in most abundance. It is thought that this habit of changing its quarters accounts for the fact that its nest is not often seen.

The Maoris adopted several methods of catching pigeons. Sometimes the fowler sat on a stage erected on a large tree, and used a contrivance consisting of a rod about five feet long, a cross-stick, and a cord, in which the feet of the pigeon were

Chatham Island Pigeons.   (*Pro. Zool. Soc.*)

caught. When the fruits of the forest were ripe, and the pigeons began to fly about in large numbers, all the men of each hapu with a reputation for strength, knowledge, and skill made "tutu," as the contrivance was called. The fowlers then sought trees with suitable tops, prepared their snares, and caught their birds. It is stated that, on some occasions, when pigeons were very plentiful, the "kill" amounted to no fewer than 200 in one day. Another method was used in connection with the birds' habit of feeding on the miro berries, which create thirst. Before the birds began to collect round the trees to eat the berries, troughs of water were placed close by, some being suspended in the trees. The birds were allowed to become accustomed to drinking from the troughs. Snares with running nooses were then placed along the edge of the troughs. The pigeons, on going to drink, placed their heads through the nooses, and were caught and strangled when they endeavoured to withdraw. The spear was also used on the pigeons.

Albino pigeons have been reported from time to time. In 1882, one was shot on the West Coast of the South Island. Its plumage was quite white, with the exception of a few blotches of brown on the back and upper part of the wings. The beak and the feet were the same colour as those of the ordinary wood pigeon. In another specimen, shot about the same time near Hokitika, part of the breast was white, as in the normal white pigeon, but the rest of the plumage was of a silver grey, light about the neck and head, and gradually darkening towards the wing-tips and tail, which was the darkest. The feathers on the back, as far as the wings, were tipped with brown, making a very even and distinct marking.

The Chatham Island Pigeon.

Hemiphaga chathamensis.

Like the last species, but the outer wing-coverts and the back are grey, and the under tail-coverts are green. Length of the wing, 10.75 in.; of the tarsus, 1.0 in. Chatham Island.

Order Galliformes.

Bill short. Legs strong, sometimes armed with a spur. Hind toe more or less developed.

Family Phasianidae.

Nostrils not hidden by feathers. Tarsi and toes naked, the hind toe raised above the ground.

Genus Coturnix.

Bill short, the nostrils covered by a scale. Wings moderate; the first quill long, the second and third longest. Tail very short, hidden by the coverts. Tarsi short. Eastern Hemisphere.

The New Zealand Quail.—Koreke.

Coturnix novae-zealandiae.

Male black, streaked with white and varied with reddish brown on the back. Chin and throat, chestnut; the breast and abdomen, spotted with white. Female browner, the chin and throat white. Eye light hazel. Length of the wing, 4.6 in.; of the tarsus, 1 in. Egg—Buff, splashed with greenish brown; length, 1.25 in. Both Islands.

There was a time when our handsome quail abounded in almost all parts of the dominion. But bush fires, cats, dogs, guns, and the onward march of civilisation soon brought disaster upon it. It is depressing to read articles, written as early as 1885, referring to the bird as a thing of the past. At one time, quail were so numerous on the large island in Lyttelton Harbour, now used as a quarantine station, that it was named "Quail Island." In the early days of Canterbury, a bag of 20 brace of quail was not looked upon as extraordinary for a day's shooting on the Plains. It was reported in the *Lyttelton Times,* in 1903, that an old settler had stated that he had shot 60 brace before breakfast on the spot where Cathedral Square, Christchurch, now stands. The slaughter in those days, the settler added, was prodigious. Mr. E. G. Wakefield, in describing Nelson in 1840, refers to long straight lines cut by the surveyors through the fern, and says: "As I walked along these future streets, quail, either singly or in convoys, frequently started up before my steps; they abound all over this part of New Zealand." In an early official report to the directors of the New Zealand Company, Colonel Wakefield, dealing with Otago, says, that "quail are plentiful over all the downs, and in the plains adjoining, and would be more so but for the hawks and kites." There is no doubt that the chief means of destruction in the southern part of the dominion was the immense prairie fires that were sent over the land in many directions.

These birds were active on the ground. It has been stated that their sense of hearing was far less acute than their sight. They often uttered a low purring sound, typical rather of an insect than of a bird. The call was given frequently in moist or wet weather. It is described as being like "twit, twit, twit, twee-twit," repeated several times in quick succession. But in very stormy, gusty weather, the birds were dull and silent, and hid away

New Zealand Quail.

among thick tussocks. In confinement, they were fond of picking about in the sand. They thrived well on soaked bread, grain of various kinds, and the larvæ of insects. The male was not an attentive mate at feeding times, and, when several were kept in the same enclosure, constant bickerings took place, though actual hostilities did not break out. The eggs required twenty-one days' incubation, and the chicks were very active as soon as they emerged from the shell. They grew rapidly, and when they had reached about four months of age they could not be easily distinguished from the adults, by contrast of either size or plumage.

The quail had a humble and lowly nest, consisting of a few bents of grass twisted into a depression of the ground. The female laid large numbers of eggs, and it is stated that as many as a dozen have been found in one nest. The birds were often observed taking a dust-bath in the sun. When flushed, they flew straight and low.

It is generally believed that this bird is extinct, but it is quite possible that it may still be found on some flats and plains that civilization has not reached.

ORDER RALLIFORMES.

Bill rather long and always hard at the tip. Wings moderate or short, rounded. Toes long, the hind toe slightly raised; the claws short.

Family Rallidae.

Nostrils in a pierced horny sheath. Toes very slender and long.

Key to the Genera.

1. Bill longer than the middle toe and claw.	2
Bill shorter than the middle toe and claw.	3
2. Bill nearly straight.	Hypotænidia.
Bill curved.	Cabalus.
3. No frontal shield.	4
A frontal shield.	6
4. Wing coverts almost reaching tips of quills.	5
Wing coverts not nearly reaching tips of quills.	Porzana.
5. Quills soft.	Ocydromus.
Quills firm.	Nesolimnas.
6. Secondaries much shorter than primaries.	Porphyrio.
Secondaries nearly as long as primaries.	Notornis.

Genus Hypotaenidia.

Bill longer than the head. Wings short, second and third quills the longest. Tail short. Tarsus shorter than the middle toe; hind toe short. Southern Asia to Australia and Polynesia.

Key to the Species.

1. Breast rufous.	H. muelleri.
Breast barred with black and white.	2
2. Back spotted with white.	H. philippensis.
Back not spotted.	H. macquariensis.

The Pectoral Rail.—MOHO-PERERU.

Hypotaenidia philippensis.

Above, brownish olive, spotted with white. Breast, abdomen, and sides, black, barred with white, and an irregular band of buff on breast. Throat and sides of the face, grey; a line of rufous through the eye to the nape. Quills dark brown banded with rufous. Eye reddish brown. Female—The buff band on the breast is very narrow. Length of the wing, 5.5 in.; of the tarsus, 1.5 in. Egg—Cream colour, spotted with dark and light chestnut and grey; length, 1.5 in. Malay Archipelago, Australia, New Zealand, and Polynesia.

Mangare Rail.

(*Cat. Birds Brit. Mus.*)

These birds live among tangled masses of grass, sedges, and rushes, which border swamps and lagoons. Owing to their shy and retiring disposition, they are seldom seen. They run through the vegetation with great agility, and never take to flight except when on the point of being captured by a dog. Even then they fly a short distance only, with a slow flapping of the wings. They live on rank weeds, and swallow sand in order to help digestion. They make no nest.

The Macquarie Island Rail.

Hypotaenidia macquariensis.

Like the last species, but the back with very few spots and the abdomen dirty white. Eye brick red. Macquarie Island.

Dieffenbach's Rail.
(Voy. Erebus and Terror.)

The Auckland Islands Rail.

Hypotaenidia muelleri.

Upper surface, reddish, brighter on the back, streaked with black; below, rufous; abdomen, black, each feather with two white bands and tipped with rufous. Length of wing, 3.3 in.; of the tarsus, 1.1 in. Auckland Islands.

Genus Cabalus.

Bill much curved. Plumage loose. Wings very short; quills soft, the fourth and fifth longest; wing coverts long. Tarsi shorter than the middle toe.

The Mangare Rail.—MATIRAKAHU.

Cabalus modestus.

Dusky brown above, greyish below, the flanks with tawny bars. Quills brown, faintly barred with tawny. Eye light brown. Young uniform brownish black. Length of the wing, 3.15 iu.; of the tarsus, 1 in. Mangare, Chatham Island Group.

Genus Nesolimnas.

Bill moderately strong, slightly curved downwards. Quills firm, wings short. Feet moderate. Chatham Islands and Lord Howe Island.

Dieffenbach's Rail.—MOERIKI.

Nesolimnas dieffenbachii.

Back, banded with buff and black. Breast, black barred with white, below which is a band of buff banded with black; under tail coverts deep ochre banded with black. Length of the wing, 4.8 in.; of the tarsus, 1.45 in. Chatham Islands. (Reported to be extinct).

Genus Ocydromus.

Bill strong, slightly curved. Wings very short, the quills soft, the secondaries and coverts lengthened. Tail soft. Tarsi strong, shorter than the middle toe. Wings armed with a spur. New Zealand only.

The Wood Hens.

Our wood hens, the wekas, have been branded as rogues and vagabonds. There is a good deal of evidence in support of the accusation, as the weka often devotes its high standard of intelligence to a very bad purpose. At the same time it has fine traits. Like common mortals, it is apt to fall from grace, but it is not utterly bad. No doubt it is often sorely tempted, and, in sitting in judgment upon it, we should make allowances and overlook its little foibles and failings, which, after all, should only endear it to us, especially as we may lose our frail friend altogether.

The charge-sheet contains several indictments. The first, and most serious, is theft. With wekas, thieving is a disease. They are confirmed kleptomaniacs. They steal not only from other birds, but also from human beings, and commit the crime, apparently, merely for the sake of stealing, as the articles they sometimes appropriate can be of no earthly use to them. Anything, from ducks' eggs to spoons, pipes, and pannikins, is good enough to be carried off. On one occasion a weka stole a silver watch from a hut in Alford Forest, and the article was only accidentally recovered a short distance from the hut. Some years ago a weka entered a bushman's hut in Peel Forest while he was away. After springing on the table, it tasted the meat, the butter, and the bread, and tumbled the remainder on the floor, with the idea, no doubt, of carrying it off. Failing in this, it took with it, so the bushman said, "one of a new pair of Sunday boots." Although the "new Sunday boot" was recovered close to the door, the bushman took revenge by killing forty wekas in less than a month.

The weka does not look like a rogue, and its appearance of childlike innocence, as it walks about with slow and deliberate steps, is deceptive. Inquisitiveness is another of its characteristics; it is of a very inquiring turn of mind, especially in regard to matters that do not concern it. Above all, it is quick tempered. "Just tread on my toes!" seems to be the idea that runs continually through its mind. When two of these fiery customers commence a quarrel, there is generally nothing for it but to fight to the finish. They sometimes go out into the open, so as to have space for sparring. When the battle is lost and won, there is woe to the vanquished, as the victor is relentless in the persecution of its fallen foe.

They possess wing-spurs, which seem to have been acquired for defensive purposes alone. Mr. W. W. Smith, in describing their fighting capabilities, says that when the birds are fighting and facing each other, the wings are elevated or arched over the back, the neck is drawn in under cover of the wings, and the spurs are pointed forwards. During a combat, they injure most the back of the head. If examined afterwards, the wounds are found to have been inflicted on only the head and neck. Along with the bruises produced by the hard bill, there are punctures caused by the wing-spurs. The punctures are always more numerous at the base of the bill, and about the eyes, and rarely extend down the neck. The effects of these bouts are noticeable for several days, the little spitfires going about with hard and swollen heads and stiff necks, carried well forward. When plenty of cold water is available, however, they bathe their heads, and, rapidly making a recovery, are again eager for the fray. The females, it is stated, do not use their spurs very much, one of the parties to the quarrel generally running away, hotly pursued by its assailant. The chase is often kept up till both are exhausted. The males, being more pugnacious, do not run so readily. Though wekas are flightless, they are remarkably quick on their legs. If pursued, they turn and double rapidly, taking advantage of every shelter until a proper refuge is found.

They are not easily flustered, as a rule. After being chased, they will emerge from a place of concealment with a look of utter unconcernedness. Captain Cook was struck with their boldness when he saw some at Dusky Sound on his second voyage. ''They inhabit the skirts of the woods and feed on the sea-beach,'' he says ''and were so tame, or foolish, as to stand and stare at us till we knocked them down with a stick.'' They do not hesitate to attack other birds and cats and rats. A weka has been known to kill a Spanish chicken at one blow. A story is related of a prospector at Lake Brunner, who kept a weka and a rat as pets. They went together every evening at tea-time to get their share of food, and sometimes they quarrelled over it. At last the weka gave the rat a decisive peck on the head, and he tumbled over dead.

The boldness and inquisitiveness of the wekas are commented on by Lady Barker in her book entitled *Station Life in New Zealand*. She says:—''I lay back on a bed of fern, watching the numbers of little birds around us. They boldly picked up our crumbs, without a thought of possible danger. Presently I felt a tug at the shawl on which I was lying. I was too lazy and dreamy to turn my head, so the next thing was a sharp dig on my arm, which hurt me dreadfully. I looked round, and there was a weka, bent on investigating the intruder into its domain. The bird looked so cool and unconcerned that I had not the heart to follow my first impulse and throw my stick at it; but my forbearance was presently rewarded with a stab on the ankle, which fairly made me jump up with a scream, when my persecutor glided gracefully away among the bushes, leaving me, like Lord Ullin, lamenting.''

These birds are omnivorous and inclined to gluttony. They have a peculiar cry, and sometimes go about at night singing a kind of duet. One call, which is more incessant than the ordinary one, and is repeated at shorter intervals, indicates rain, and is an excellent barometer. The ordinary call is generally led off by the female, and is answered in all directions by both sexes. This, however, it is stated, is peculiar to paired birds before or after mating, as the male or female is often heard solitary, answering others in the distance, while its mate is on the nest.

In speaking of the weka's good qualities, Mr. W. W. Smith denounces the ''ignorant and mistaken prejudice'' shown towards this remarkable bird. He finds that too much cannot be said in its favour, and believes that all prejudice would be overcome if proper attention were given to its habits, as ''the mere destruction of a few eggs in or near the poultry yard, or the disturbing of a few pheasants in reserves, may be overlooked, when it is remembered that the weka renders inestimable services in destroying vermin.'' He adds that its struggle for existence is greater than that of any other native bird. ''This is owing to thousands perishing annually in the fires that swept over large areas of tussock lands, to merciless destruction by dog and gun, and, above all, to stoats and weasels. To their credit it must be stated that they are of great service to the squatters and farmers in consuming the larvæ of odontria, which devastate lawns and English grass paddocks. When a brood is hatched near paddocks infested with these grubs, the parent birds lead them there, and dig vigorously over the ground, rooting them out with their

powerful bills to feed the young. If encouraged about homesteads, they are heard during the night tapping on the walls of
dwellings and outhouses, pecking off the spiders and insects
secreted there. When bags or sheepskins are found lying on the
ground, they drag them away or turn them over to procure the
worms, beetles, or woodlice hidden beneath. Occasionally, during
their nocturnal rambles, they discover the carcase of a sheep, and
commence pulling off the wool until they effect an opening in the
flesh. Here they fare sumptuously for weeks, often secreting
themselves in the nearest cover, and returning night after night

Nest of South Island Wood Hen.

to feed on the carcase. They are also death to rats and mice, and
help in the destruction of young rabbits. When enclosed in small
yards, they become tamer than domestic fowls, thrusting their
heads through the meshes of the wire and feeding from the
hand.''
    Mr. Smith, in 1885, endeavoured to procure hybrids between
the weka and the domestic game fowls, and so settle the question
of crossing. He raised a nest of young wekas with one domestic
well-bred game cock. He was not successful in attaining his
object, and indeed it is almost impossible that birds belonging to
different orders should breed together. One bird laid for the first

time on August 4th, and by August 20th it had produced eleven eggs. It then discontinued laying for nine days. After that it continued to lay, on an average, every two days, missing a day or two occasionally. Two of them contracted the pernicious practice of eating their own eggs. The eggs, when cooked, were found to be slightly inferior to those of the domestic fowl.

He states that the weka, in its natural state, pairs only once, and remains permanently paired, unless by some means the pair are separated. Writing in May, 1903, he says that, during the past three years, he renewed his attempts to breed hybrids, but still without success. By carefully feeding the wekas with suitable food, he had no difficulty in getting them to lay from two to two and a half dozen eggs each in a season. Before laying, they generally utter a peculiar call, and then they must be shut off until they lay.

The nest of the weka is placed in different situations. Mr. Potts has found it in a tuft of celmisia, a grass tussock, a thicket of young plants, and the outskirts of a bush, and under the shelter of a rock, without any attempt at concealment. The nest is large, and the inside is shaped like a basin. The principal material used is generally grass. There are from five to seven eggs. The young may be seen, like chickens, following the old bird, who collects them around her with a call of "toom toom" repeated quickly, and much lower in tone than the booming note to which the weka sometimes gives utterance. As the young grow up, the dark brown of their early days gives place to a more mottled plumage when they are about one-third grown. Although the legs become lighter in colour, the beak still retains its dark appearance.

Key to the Species.

1. Tail black with no cross bars.	2
Some of the tail feathers barred with rufous.	3
2. Dark brown.	O. brachypterus.
Rusty brown above, grey below.	O. earli.
3. Middle tail feathers brownish black.	O. finschi.
Middle tail feathers banded.	4
4. Larger, tarsus 2.3.	O. hectori.
Smaller, tarsus 2.	O. australis.

The North Island Wood Hen.—WEKA.

Ocydromus earli.

Above brownish rufous, streaked with brownish black. Below grey, the breast tinged with rufous. Quills banded with black and rufous. Tail without mark. Eye reddish brown. Length of the wing, 7.75 in.; of the tarsus, 2.4 in. The female is smaller than the male. Egg— Pinkish white, with reddish and violet spots sparingly distributed over the surface; length, 2.4 in. North Island and Stewart Island.

The Black Wood Hen.

Ocydromus brachypterus.

Black, each feather margined with reddish brown. Throat, sides of the face, and abdomen dark grey. Quills sparingly marked with rufous on the inner webs only. Eye reddish brown. Tail without mark. Length of the wing, 7 in.; of the tarsus, 2.2 in. The female is smaller than the male. West Coast Sounds of the South Island.

Mr. Reischek specially observed the black wood hens during his visit to the West Coast Sounds in 1884. He saw them mostly at dusk, roaming about stony river beds, seeking food. The numerous dead trees that are swept down along the banks by floods afforded them hiding places. He also saw them on the sea shore, picking mussels and crabs, and on the mountains, as high as 2000 feet above sea level, but there they were scarce. During the day they concealed themselves under roots and in hollow trees, their hiding places having generally two or three entrances, so that, in case of disturbance, they could easily escape. He was amused once at seeing a dog digging vigilantly at a burrow, while the wood hen was quietly stealing away. On the dog pursuing her, she dodged him in the coolest manner for nearly a quarter-of-an-hour, by going under trees, and always taking care to keep on the opposite side from that on which he was. But, on Mr. Reischek going to the dog's assistance, she gave a shrill whistle, and ran swiftly away.

When undisturbed, he says, these birds are very bold and tame. At Dusky Sound, a shining black wood hen went every morning and evening to his camp in the gorge, uttering a shrill whistle of one note, and, on his throwing her a piece of biscuit, she would pick it up, throw it on the ground till it broke, and then eat it.

She became so tame that she would walk round his dog, and come into the tent, and, on a second visit to the camp, he found that she still haunted the place. On one occasion, at daylight, he was awakened by a noise, and, on looking up, saw one of these birds amusing itself with his slippers, but, on his moving, she retired.

The breeding season is in January, when the birds lay from two to three eggs. Mr. Reischek saw in April two females, with three young birds each, fully feathered. The young were duller in plumage, and smaller in size than the parents. The male and female do not differ in plumage, but there is a slight difference in size, the latter being the smaller. These birds vary much in plumage, but the jet black ones are rare.

(Voy. Erebus and Terror.)

South Island Wood Hen.

Finsch's Wood Hen.

Ocydromus finschi.

Brownish black, with spots of buff on the outer margins of each feather. Throat, abdomen, and thighs dark grey. Quills banded on both webs with dull rufous. Middle tail feathers brownish black, the outer ones with spots of buff on the margins of the webs. Eye reddish brown. Length of the wing, 7 in.; of the tarsus, 2.3 in. Te Anau district, South Island.

The South Island Wood Hen.—Weka.

Ocydromus australis.

Rusty red, streaked with brownish black. Below grey, the breast tinged with ferruginous. Quills and middle tail feathers barred with rufous. Eye reddish brown. Length of the wing, 6.7 in.; of the tarsus, 2 in. South Island and Stewart Island. The egg is figured in the *Proceedings* of the Zoological Society of London for 1852, plate Aves XLVI. It is pinkish white, with reddish and violet spots sparingly distributed over the surface; length, 2.85 in.

The Hill Wood Hen.

Ocydromus hectori.

Olivaceous, or fawn, streaked with brownish black. Below grey, tinged with olivaceous or fawn on the breast. Middle tail feathers generally with a black streak down the shaft, and rufous margins. Eye reddish brown. Length of the wing, 7.75 in.; of the tarsus, 4.75 in. The female is smaller than the male. Egg—Pinkish white, with red and purple markings sparingly distributed over the surface; length, 2.86 in. South Island, in sub-alpine districts.

Genus Porzana.

Bill shorter than the middle toe with claw. Wings short, the second and third quills the longest; the secondaries considerably shorter than the primaries. Tarsus shorter than the middle toe with claw. Found over nearly the whole world.

The Marsh Rail.—Koitareke or Koreke.

Porzana affinis.

Above, brown, spotted with white and varied with black. Abdomen barred with black and white. Eye dull red. Length of the wing, 3.3 in.; of the tarsus, 1 in. Egg—Olive brown, polished; length, 0.85 in. Both Islands and the Chatham Islands.

These solitary little birds live in flax swamps and hide themselves with much dexterity. They might live a long time in a place without their presence being suspected, were it not that unlucky individuals are occasionally brought in by a prowling cat. The bird is closely allied to the spotted water crake of Australia.

The Swamp Rail.—Putoto or Puweto.

Porzana tabuensis.

Slate blue, brownish on the back.   Under tail coverts, black banded
with white. Eye bright red. Length of the wing, 3 in.; of the tarsus,
1 in. In the young the throat and part of the breast are white. Egg—
Pale creamy brown, freckled with darker; length, 1.3 in. Both Islands
and the Chatham Islands.   Also Polynesia, New Caledonia, Australia,
and the Philippine Islands.

Marsh Rail.   Swamp Rail.   *(Buller.)*

The swamp rail also frequents swampy places, especially raupo
swamps. It is rarely seen, but an observer, by patiently standing
still, may hear its low purring note close by, but in what direction
it is difficult to say. According to Mr. Gould, it swims with
grace and elegance, sporting with ease among the floating leaves
of aquatic plants in search of water snails, which form its
principal food.

Genus Porphyrio.

Bill short, elevated at the base, which is flat and dilated on the forehead. Second to the fourth quills longest. Toes very long, free at the base, the hind toe long. The warmer parts of the eastern hemisphere.

Swamp Hen.   *(Buller)*

The Swamp Hen.—PUKEKO or PAKURA.

Porphyrio melanonotus.

Above, deep black; breast, indigo blue; abdomen, black; under tail coverts, white. Eye crimson. Length of the wing, 10 in.; of the tarsus, 3.6 in. Egg—Greyish brown, with dots and blotches of grey and brownish violet; length, 2.2 in. Both Islands. Also Australia, Norfolk Island, Lord Howe Island, and New Guinea.

The swamp hen is generally known by its Maori name, "pukeko." It is a handsome and graceful rail. Its nest of grass is built in swamps, and is sometimes found quite surrounded by water. As a rule, about five eggs are laid in the nest, but the number varies considerably. The young run about as soon as they are hatched, and, when disturbed, conceal themselves with much art. They are thickly clothed with black, velvety down, interspersed with fine hair-like points of silver grey. The legs are dullish red, and the beak has a yellowish ivory appearance, which contrasts with the colour of the body. When feeding, the pukeko often lifts the food to its mouth with its claw, as a parrot does. Its flesh is very palatable, and the bird is therefore the object of many shooting expeditions. It feeds chiefly in the early morning and in the evening, hiding itself away among the flax bushes in the day time. Vegetable substances form its principal food, but it also eats insects. The pukeko is easily tamed, and, when kept with domestic fowls, makes friends with them.

The Chatham Island Swamp Hen.

Porphyrio chathamensis.

Like the last, but the breast is bright blue.   Length of the wing, 10.3 in.; of the tarsus, 3.7 in.   Chatham Islands.

Genus Notornis.

Bill short and deep, compressed, the frontal shield produced backwards over the eye. Wings short, the quills soft, the third to the seventh equal and longest; the wing coverts much elongated, nearly hiding the quills, which are not much longer than the secondaries. Legs and feet strong; the tarsus as long as the middle toe with claw. Hind toe short and elevated; the claws hooked. New Zealand and Norfolk Island.

Takahe.*

Notornis hochstetteri.

Above olive green, with some blue shading. Wings blackish blue. Head, neck, and lower surface dark purplish blue; under tail-coverts white. Length of the wing, 9 in.; of the tarsus, 3.7 in. South-west portion of the South Island.

---

*Mr. James Cowan states that the Maoris of the South say that this word is "TAKAHEA."

A glamour of romance has been shed round Notornis hoch
stetteri, on account of its strange and isolated position. When
fossil remains of the genus were first brought to light in New
Zealand, naturalists thought that all the representatives here had

(*Buller.*)

Takahe.

passed away with the extinct moa, eagle, and swan. Two years
afterwards, however, a live specimen was found, and it was then
known that the name Notornis could be added to the list of notable
living birds of New Zealand. The circumstances surrounding its
discovery, its great rarity, and the solitary life it leads in the
fastnesses of the West Coast Sounds, add to the interest this
bird excites. Only four specimens have been obtained. Two are

in the British Museum, one in the Dresden Museum, and the other in the Dunedin Museum, Otago.

It is a large, heavy, and flightless rail, with massive bill and legs, a very handsome and striking plumage, and a solemn cast of countenance. The bill and legs, in contrast to the prevailing colours of the plumage, are bright red. Mr. Gould says that it might be mistaken for a gigantic kind of Porphyrio, but that an examination of its structure shows it to be generically distinct. In the form of its bill and general colouring it is allied to Porphyrio, and in the structure of its feet to Tribonyx, a rail which is found in Australia and Tasmania, and which, though it rarely resorts to flight, can run well. In the feebleness of its wings and the structure of its tail, however, it differs from both. Having personally observed the habits of Tribonyx and Porphyrio, Mr. Gould affirms that the habits and economy of Notornis more closely resemble those of the former than those of the latter. Being deprived by the feeble structure of its wings of the power of flight, it is compelled to depend on its swiftness of foot for the means of evading its natural enemies, so that, as with Tribonyx, a very shy bird, a person may be in its vicinity for weeks without catching a glimpse of it.

From the thickness of its plumage, and the great length of its back feathers, Mr. Gould infers that it affects low and humid situations, marshes, the banks of rivers, and the coverts of dripping ferns, and, like the Porphyrio, doubtless enjoys the power of swimming, but would seem, from the structure of its legs, to be more terrestrial in its habits than the members of that genus.

The first discovery was made by Mr. W. Mantell, in 1847. In a bed of volcanic ashes at Waingongoro, in the North Island, he found a number of fossil moa bones, which he sent to Sir Richard Owen. "I detected among them," Sir Richard says, "portions of the skull of a bird about the size of a turkey." The back part of the head was broad and sloping, something like that of the moa, but there were differences that induced him to make an extensive series of comparisons. These brought him ultimately to the family of the coots and water hens, where he found the greatest number of cranial similarities with the fossils. Though the evidence upon which he worked was scanty and incomplete, being merely portions of a fossilised skull, his knowledge and wonderfully accurate observations convinced him that there had existed in New Zealand a bird which might be deemed to be the giant of the coots, with so close an affinity to Porphyrio as at most to suggest only a sub-generic distinction. So he thereupon established a genus, named it Notornis, and gave to this colony an important addition to its avi-fauna; the species he rightly called mantelli, after the discoverer. For some time the living species was also called mantelli, but Dr. Meyer, after examining a skeleton in the Dresden Museum, concluded that the South Island bird was a different species, and he named it hochstetteri.

Strangely enough, it was Mr. Mantell who obtained the skin of the first recorded live specimen of Notornis made known to science. Dr. Mantell has described how the bird came into his son's possession. In 1849, it was taken by some sealers who were pursuing their vocation in Dusky Bay. At Duck Cove, Resolution Island, they saw the trail of a large and unknown bird on the snow, with which the ground was then covered. They followed the footprints until they obtained a sight of the Notornis. Their dogs at once pursued it, and, after a long chase, caught it alive in the gully of a sound behind Resolution Island. It ran with great speed, and, on being captured, uttered loud screams, and fought and struggled violently. It was kept alive for three days on board the schooner, and was then killed. The precious body was roasted and eaten by the crew, each partaking of the dainty, which was declared to be delicious. Mr. Mantell secured the skin. He states that, according to native traditions, a large rail was contemporary with the moa, and formed an article of food among the Maoris' ancestors. It was known to the North Islanders by the name of moho, and to the South Islanders as takahe; but, he adds, the bird was considered by both natives and Europeans to have been long before exterminated by the wild cats and dogs, an individual not having been seen or heard of since the arrival of the English colonists. To the natives of the pas or villages on his homeward route, and at Wellington, the bird was a great novelty and excited much interest.

"I may add," Dr. Mantell says, in a paper he read before the Zoological Society on November 12th, 1850, "that upon comparing the head of the bird with the fossil cranium and mandibles, and the figures and descriptions in the *Zoological Transactions*, my son was at once convinced of their identity; and so delighted was he by the discovery of a living example of one of the supposed extinct contemporaries of the moa that he immediately wrote to me, and mentioned that the skull and beak were alike in the recent and fossil specimens, and that the abbreviated and feeble development of the wings, both in their bones and plumage, were in perfect accordance with the indications afforded by the fossil humerus and sternum found by him at Waingongoro, and now in the British Museum." Shortly afterwards a second specimen was caught by some Maoris on Secretary Island. This also was obtained by Mr. Mantell, and both were sent to the British Museum, where they remain.

Thirty years passed, and nothing further was seen or heard of the Notornis. The general belief was that the bird had really become quite extinct, and value was added to the specimens secured.

But in 1879 the skin and skeleton of a bird caught on the Te Anau Downs, between the Mararoa and Upokororo Rivers, by some men while on a rabbiting expedition, were reported. They were sent to Dunedin, forwarded to England, and sold to the Dresden Museum for £110. The specimen attracted a great deal of interest in Europe, and was the subject of papers read before the Zoological Society in London by Sir R. Owen and Professor Newton.

The fourth specimen was caught by Messrs. D. and J. Ross, also at Lake Te Anau, in 1898. On August 7th, while lying awake in their bunks, they heard a bird-call that struck them as being unusual. It came from the bush near the edge of the lake, and about a hundred yards from their camp. They thought that the sound was not unlike a double call which is often made by the Californian quail, but was not so sharp and clear. In the evening, just before darkness set in, one of the brothers, while walking along the beach, saw his dog disappear in the bush, and emerge shortly after with a bird in its mouth. The bird was not dead, and it was immediately taken to the camp, where it expired shortly after its capture. A Notornis had been found. The captors immediately took boat to the foot of the lake, twenty-five miles away. Early in the morning they reached their destination with their precious prize, and had to convey it to Lumsden and on to Invercargill, from which place it was forwarded to Dunedin, and placed in the Museum. Later on, it was the subject of a question in the House of Representatives by Mr. R. McNab, the member for Mataura, and was purchased by the Government for £300.

This bird proved to be a young female, in a thoroughly healthy and clean condition. Professor Benham took it in hand at Dunedin. He is the first naturalist to examine the internal anatomy of Notornis. His observations have been communicated to the Zoological Society of London, and have also been published in the *Transactions* of the New Zealand Institute, Volume XXXI., page 151, together with plates.

He states that an interesting fact in regard to the colour of the bird, which must be of great value to it, is immediately noticeable in examining the skin in different lights. He says:—"The best effect is obtained when the light and the eye are in the same direction, and if the front of the bird be looked at. But if we now look at the back of the bird, as it would be seen if it were running away from the pursuer, no bright tint is seen. The colour is dull, dirty grey, admirably adapted for concealing the bird as it escapes into the bush or amongst any growth higher than itself, and capable of casting a shade. The white under-coverts of the tail form, however, a conspicuous mark in the bird, as in so many of its allies, and though more noticeable when seen from the side in contrast with the brighter colours of back and wings, yet, from behind, the white is not so noticeable as might be imagined. It is difficult to say what meaning is to be attributed to this white tail. In many cases, as with antelopes and rabbits, it is a 'recognition mark,' as Wallace has called it, enabling members of a herd to find their fellows at night, or to follow the lead of others in escaping enemies. It usually occurs in animals of gregarious habits, and we should judge therefrom that Notornis is gregarious. It is all the more curious, then, that isolated individuals should have been caught, and nothing seen of their fellows. But from what enemy does the Notornis flee? What native animal of the present day preys on Notornis? Probably none. Then this 'recognition mark' must have come down from a time and place in which there were enemies.''

ORDER HERODIONES.

Large birds· with long legs, neck, and bill. Toes long and slender, the hind toe not elevated above the others. Part of the head and neck naked.

Family Ardeidae.

Bill long, strong, and acute. Wings rounded. Tail short. Outer toe with a broad basal web; middle claw pectinated. Space between the mouth and the eyes bare.

Key to the Genera.

1. Tail feathers ten.	2
Tail feathers twelve.	3
2. Tarsus equal to middle toe.	Ardetta.
Tarsus shorter than middle toe.	Botaurus.
3. Bare portion of tibia is less than inner toe and claw.	Demiegretta.
Bare portion of tibia at least equal to inner toe and claw.	4
4. Bill not longer than middle toe and claw.	Herodias
Bill longer than middle toe and claw.	Notophoyx.

Genus Herodias.

Bill not serrated, but a distinct notch at the end of the upper mandible; not longer than the middle toe and claw. Tail feathers twelve. No crest, but a well developed dorsal train. Nearly cosmopolitan.

The White Heron.—Kотuки.

Herodias timoriensis.

White; the bill yellow; naked space before and behind the eye, greenish yellow. Legs below the knee and toes, black. Eye yellow. Length of the wing 16.17 inches; of the tarsus, 6.25 inches. Egg— Pale green; length 2.2 inches. From China and Japan, through the Malay Archipelago to Australia and New Zealand; also parts of Africa and Europe.

According to a Maori proverb, the white heron has been a very rare bird in New Zealand for a long time. This may have been the case in the North Island, but certainly not in the South. When the first settlers came to the South Island, this bird was not uncommon on the margins of lakes and rivers. Lake Heron, in the Ashburton County, was named by Mr. Potts, in 1857, on account of the number of these birds which frequented its shores. But the white heron is now very rare in New Zealand. It has fallen a prey to the gunner, who has killed it, not for food, but in a spirit of wanton destructiveness.

There is a white heronry at Okarito, Westland, which was visited by Mr. Potts and his sons in December, 1871. He counted twenty nests in trees in close association with the nests of the white-throated shag. The nests were firmly made with sticks, and contained three or four pale green eggs. These birds are still seen occasionally in the Okarito district. The female commences laying in the third week in November, and incubates for four weeks. The young birds remain in the nest for a considerable time. In the breeding season both sexes develop beautiful white dorsal plumes, well known to milliners by the name of "egrets."

The white heron feeds principally on fish, which it catches by patiently waiting until they come near. "It is a sight for the naturalist to remember when his eyes fall upon a kotuku, silently standing with meditative mien in some shallow pool awaiting its prey, ready for the fatal dart," says Mr. Potts. "Its spotless plumage, thrown into bold relief against the backing of a mass of foliage, is mirrored distortedly by the rippling water. Long is the patient watch maintained in stilly silence. At length the glistening prey glides unwarily within reach of the spear-like bill. One quick stroke, almost too swift for the eye to follow, and a slight movement of the neck, tell that the prey is captured and engulfed; and the silent watch is once more resumed.''

When flying, the head is kept far back, the tip of the bill being behind the forward curve of the neck; and the legs are stretched out straight behind. The large rounded wings beat slowly, but the progress of the bird is rapid.

The Maoris placed a high value on the plumes of the white heron. Mr. Elsdon Best states that among the people who lived in the fastnesses of Tuhoe Land, and also probably in other places, the plumes were tapu (sacred). If a man wore one while eating, no woman might join in the meal unless the wearer took it off and placed it on one side. It was supposed that if a woman persisted in joining in the meal, she would become bald.

Genus Notophoyx.

Bill longer than the middle toe and claw. Tarsus moderate, not twice the length of the outer toe and claw. Tail with twelve feathers; dorsal train not extending beyond the tail; a full crest, but no nape plumes. Australia, New Caledonia, the Moluccas, and Cook Islands.

The White-fronted Heron.—MATUKU.
Notophoyx novae-hollandiae.

Bluish grey; throat, forehead and over the eye, white; breast tinged with pink; eye yellow. Length of the wing, 12 in.; of the tarsus, 3.5 in. Egg—Bluish green; length, 1.9 in. The same range as that of the genus.

The white-fronted heron, according to Mr. Potts, used to be far from common in Canterbury. In Australia it builds on trees and makes its nest of sticks.

(Buller.)

Blue Heron.   White-fronted Heron.

Genus Demiegretta.

Bill not serrated, but the upper mandible notched near the tip, longer than the tarsus; the latter longer than the middle toe. Tail of twelve feathers. Eastern Asia to Australia and Polynesia.

The Blue Heron.—MATUKU-MOANA.

Demiegretta sacra.

Dark slate grey; brownish on the wing coverts; chin and throat, white; legs and feet, yellowish green; eye, yellow. Length of the wing, 11 in.; of the tarsus, 3 in. Egg—Pale greenish blue; length 1.7 to 1.9 in. The same range as the genus.

This is a coast bird. It is much more common in the north than in the south. Mr. Potts says that it ''may be observed alighting on reefs immediately on the ebb of the tide, so that the birds stand in the water. I have noticed this at Kawau and elsewhere. I have an egg from near the Kidnappers: it was taken from a nest of sticks and debris in a recess among rocks, just above the sea. With us, in the south, this bird is of rare occurrence. Occasionally it may be observed on the flats at the head of one of the bays, but we only look upon it as a visitor. When we have noticed it on the wing, it had been flying low, just skirting the shore, with deliberate, almost heavy, flight.''

Genus Ardetta.

Bill serrated. Legs feathered down to the tarsal joint. Middle toe and claw about equal to the tarsus. Tail feathers, ten. The area behind the eye feathered. Nearly cosmopolitan.

The Little Bittern.—KAORIKI.

Ardetta pusilla.

Back and quills, dark brown; top of the head, greenish black; front of the neck, buff, passing into chestnut towards the back of the neck; a stripe of chestnut, streaked with brown, down the front of the neck. Wing coverts, buff, striped with dark brown; abdomen, buffy white, streaked with grey. In the young the back is varied with rufous, and some of the primaries and secondaries are tipped with the same colour. Eye orange. Length of the wing, 6 in.; of the tarsus, 1.75 in. Australia and New Zealand.

Only three or four specimens of the little bittern have been obtained in Westland. It is said to be so quiet in its habits that it will remain still when approached, and will almost allow itself to be caught by hand. According to Dr. Docherty, these birds are found on salt water lagoons on the sea shore, always hugging the timber-lined side. "I have seen them in two positions." he adds, "standing on the banks of the lagoon with their heads bent forward, studiously watching the water, and standing straight up, almost perpendicular. They live on small fishes. They are very solitary, are always found alone, and stand for hours in one place. They breed on the ground, in very obscure places, and are, on the whole, rare birds. I have never heard their cry."

Genus Botaurus.

Bill serrated, about equal in length to the inner toe with claw. Middle toe and claw much longer than the tarsus; hind claw very long, nearly as long as the toe. Nearly cosmopolitan.

The Bittern.—MATUKU-HUREPO.

Botaurus poeciloptilus.

Blackish brown, varied with buff. Length of the wing, 13 to 14 in.; of the tarsus, 4 in. Eye, yellow. Egg—Brownish olive; length, 2 in. Australia, New Caledonia, and New Zealand.

The bittern is not as common as it used to be. The scarcity is attributed to the draining of many swamps; but it is still often seen in suitable localities. Mr. Potts says that it has an extended breeding season. The nest is built of raupo or other aquatic plants; it is flat on the top, and, when built in the water, stands about six inches above the surface. The eggs are four in number, and are small for the size of the bird. The bittern makes a loud booming noise, which may be heard for a long distance at night, and which sounds something like the roaring of a bull a long way off. It has a curious habit of standing for a long time with its bill pointing straight up to the sky, as if it were studying the weather.

In a communication to the authors, Mr. C. Lewis, of Halswell, says:—"You say, 'it has a curious habit of standing for a long

time with its bill pointing up to the sky, as if it were studying the
weather.' This I have never noted, but its habit of lying or
squatting down with its bill pointing straight up is, I believe,
assumed for protective purposes. The bittern frequents marshy

Bittern.   (*Meyer.*)

ground, which, especially when covered with water, gives ample
notice of approaching footsteps. Frequently, on coming round a
patch of cover, I have seen what looked like a dried stump of
gorse, or a wisp of dried raupo, flax, or rushes, but on closer
examination it proved to be a bittern sitting motionless with its
bill pointing upwards. On one occasion, I saw a bittern walking

about near the edge of a swamp, and I successfully stalked him. On emerging from behind the rushes which had covered my approach, all I could see was the 'gorse stump' before mentioned. Realising the position, I watched it for several minutes, during which I could detect no movement. But it was the bittern nevertheless. Presumably he had heard my approaching footsteps."

ORDER LIMICOLAE.

Bill longer than the head, which has no bare skin. Primaries eleven, the fifth secondary wanting. Legs usually long, the toes partially webbed or not webbed. Hind toe either absent or elevated above the others.

Key to the Families.

Nasal grooves not extending beyond half the length of the bill.	Charadriidæ.
Nasal grooves extending along the greater part of the length of the bill.	Scolopacidæ.

Family Charadriidae.

Bill short or moderate, often swollen at tip. Hind toe either absent or small and slender.

Key to the Genera

1. Bill not swollen at the tip.	2
Bill swollen at the tip.	5
2. Legs very long, tarsus more than twice the middle toe.	3
Tarsus less than twice the middle toe and claw.	4
3. Bill straight.	Himantopus.
Bill curved upwards.	Recurvirostra.
4. Bill short, a hind toe.	Arenaria.
Bill long, no hind toe.	Hæmatopus.
5. Bill shorter than middle toe.	Charadrius.
Bill equal to middle toe.	Ochthodromus.
Bill longer than middle toe.	6
6. Bill straight.	Thinornis.
Bill bent to the right.	Anarhynchus.

Genus Arenaria.

Bill not longer than the head. First quill longest. Tarsus as long as the middle toe, reticulated behind, transversely scaled in front. Hind toe present. All parts of the world.

The Turnstone.

Arenaria interpres.

Winter plumage—Above blackish brown; throat, abdomen, and over the tail white, breast brownish black; tail white, with a broad band near tip. Bill black, legs red, shafts of the quills white; eye blackish brown. In the summer the head and lower surface are white, the head and breast mottled with black. Length of the wing, 6 in.; of the tarsus, 1 in. Migratory.

(Meyer.)
Breeding plumage of Turnstone.

The turnstone breeds in the Northern Hemisphere, but not so far north as the godwit, and it is found in its breeding-dress in India and Ceylon. In early autumn it leaves its more northern breeding-grounds, and some pass through the Malay Archipelago and New Guinea to Australia and Tasmania. In New Zealand it arrives in November, and leaves in March or April, almost all the birds being in winter plumage. A few remain, however, and take on their summer plumage, although they have never been known to breed here. Stragglers occasionally spread from Fiji through eastern Polynesia, but there is no regular migration eastward of Fiji.

Genus Haematopus.

Bill longer than the head, strong, rather concave upwards, much compressed at the tip. First quill the longest. Tarsi longer than the middle toe. Almost cosmopolitan.

The Oyster Catcher.—TOREA.

Haematopus longirostris.

Above greenish black; below and over the tail white; bill and legs crimson; a white band on the wing; eye crimson. Length of the wing, 11 in.; of the tarsus, 2 in. Egg—Pale yellow brown, spotted and blotched with brownish black; length, 2.25 in. Molucca Islands, Australia, New Zealand, and the Chatham Islands.

Oyster Catcher.

(*Meyer.*)

An interesting description of this bird has been given by Mr. Potts, as follows:—"Years ago, in 1858, before the shores and estuaries were frequented so much, these shore birds were exceedingly abundant. Thousands of them were to be seen together, and, as late as 1871, they were abundant on the mud-flats in Lyttelton Harbour. The oyster catcher is one of the wariest and most restless of our birds. It is always ready with its clamorous alarm-note to wake up each echo and disturb every bird within sound of its shrill cry. But, in the breeding season, it exhibits an intensity of slyness that is almost supernatural. Usually it breeds in our river-beds, on the sandy spits, without any other shelter than what may be afforded by some drift flax, grass, or stick, near which it makes, or discovers, a slight depression in which to deposit its eggs. These are usually three in number. The young are grey, with a dark longitudinal stripe on each side above the wing. They are very active, and are early led by the old birds to the margins of the water-holes or pools. On being alarmed, the old bird sidles off the nest quietly, takes advantage of any broken ground that helps to conceal its movements from observation, and makes a long detour. A close scrutiny will very frequently enable the observer to detect the head of the bird carefully peering out behind some vantage ground watching all his proceedings.''

In the winter the oyster catchers assemble in large flocks on the coast, but, when the breeding season comes on, they retire in pairs up the river-beds, sometimes far into the mountains.

The Red Bill.—TOREA-PANGO.

Haematopus unicolor.

Greenish black; bill and legs crimson; eye crimson. Length of the wing, 10.5 in.; of the tarsus, 2 in. Egg—Greenish brown, thickly spotted with black and brown; length, 2.45 in. Australia and New Zealand.

The red-bill is rarer than the oyster catcher. It is found chiefly in the sounds and inlets of the west coast of the South Island. Its food is molluscs and crustaceans, left on the sea shore among the tangled masses of brown sea-weed. It generally breeds near the sea, and only occasionally in river-beds, its nest being a few twigs and grass-culms. The eggs are either two or three in number. The young are very active, and run about freely with the parent birds   Like many of the wading tribes that share the same gift, this bird, it is stated, displays wonderful capacity for taking care of its offspring, espying danger from afar, warning its young, and instantly adopting schemes for misleading intruders. In these attempts at preservation the young ones second the efforts of the parents with remarkable courage, promptitude, and decision.

(*Voy. Erebus and Terror.*)
Red Bill.

Genus Charadrius.

Both mandibles grooved, that on the upper one extending for two-thirds of its length. No facial wattle. Wings long and pointed, first quill the longest; no wing spur. Inner secondaries very long and pointed. Hind toe wanting. All over the world.

The Spotted Plover.

Charadrius dominicus.

Winter plumage—Above, blackish brown, spotted with yellow or yellowish white. Below, yellowish white. Eye, dark brown. Length of the wing, 6.5 inches; of the tarsus, 1.7 inch. In the breeding plumage, the throat and the middle of the lower surface are black. Migratory.

The spotted plover is another Siberian bird, which migrates regularly to Australia, has spread over Polynesia, and, according to Dr. Graffe, has become a resident at Tongatabu. In the

Northern Hemisphere it takes on its summer plumage in April, and changes into the winter plumage in August or September. It is a common bird in New Caledonia and Fiji, and Mr. E. L. Layard says that in the former island he found on April 20th, 1877, a female followed by a couple of chicks a few days old. But he also says that the old birds attain their full breeding plumage in May, which is the same time as in the Northern Hemisphere. It would seem from this that the breeding of the

(*Meyer.*)

Spotted Plover: breeding plumage.

bird in the island was what might be called accidental, and, as the birds have not been long enough to change the time of breeding or of moulting, it is probable that all are migrants, but that some delay moving northwards until they have attained the breeding plumage.

Something of the same nature probably happens in Australia, as Mr. Gould says that the uniform black under surface, which is the complete breeding plumage, is seldom seen there; and all the specimens in the British Museum from Malay Archipelago Australia, and Polynesia, are in the winter plumage.

In New Zealand the bird is rare, having been recorded only a few times in the North Island, while for the first time in history it made its appearance in the South Island in the summer of 1900. Mr. W. W. Smith says that he has seen a good many of these birds in the Ashburton River bed. There is a specimen in the Canterbury Museum, which was shot at Lake Ellesmere, in 1900. Two specimens shot near Auckland early in December, 1880, were in winter plumage, but showed signs of being about to put on their summer dress. The specimen in the Canterbury Museum is in winter plumage, as also is one shot at the Bluff; and Mr. Smith says that the plumage of the birds in the Ashburton River bed varied but slightly. But in the Canterbury Museum there is another specimen in the full breeding plumage, which was shot on March 5th, 1902. Mr. C. H. Robson found a pair breeding at Portland Island, on January 9th, and, as he says the birds undergo little or no change of plumage from winter to summer, which is a mistake, it may be presumed that the birds he saw were also in the winter or non-breeding plumage. This is very remarkable, for with introduced European birds, such as the starling, the linnet, and the redpoll, the change of plumage goes with the breeding season, as it did in Europe; both, on coming into the Southern Hemisphere, have changed together.

We cannot, therefore, think that the birds breeding at Portland Island were true residents, for, if they had been long in New Zealand, it is probable they would have acquired their summer plumage in the breeding season.

Genus *Ochthodromus*.

Differs from Charadrius in not having a black abdomen in the summer, and in the upper surface not being spotted. Widely distributed; not in Polynesia.

The Dotterel.—Tuturiwhatu.

Ochthodromus obscurus.

Above, brown; below rufous. Forehead, chin, and under tail coverts, white. In the winter the under parts are pure white, with a band of brown on the breast. Eye, blackish brown. Length of the wing, 6.5 in.; of the tarsus, 1.6 in. Egg—Brownish yellow, spotted and blotched with black; length, 1.75 in. Both Islands.

The dotterel is another bird that has had to beat a retreat before civilisation. At one time it was found on the Canterbury Plains, but it has now gone up into the mountains of the back country. It is described as an excellent game bird, and one of its principal characteristics is the artful manner in which it endeavours to protect its young from danger. No description, Mr. Potts says, can do justice to its contrivances for misleading an intruder. Swift runs, short flights hither and thither, with the click, clicking call resounding again and again, often effectually puzzle the disappointed collector, and lead him astray Some well-grassed land, in a situation that affords plenty of insect life, such as crane-flies and grasshoppers, is generally selected as a fitting place for a nest. The structure is very slight, and may easily escape observation. A few grass bents are twisted into a rounded shape in a slight hollow in the ground, and the whole is put together so loosely that it is difficult to pick it up and at the same time preserve its form. There are three eggs, and they fill the nest. The young leave the nest almost immediately after hatching, and accompany the parents on their rambles in search of food.

The Banded Dotterel.—Pohowera.

Ochthodromus bicinctus.

Above, greyish brown; the forehead, white, margined above and below with black; a black line from the gape through the eye down the side of the neck. Below, white, with a band of black on the breast, and another of chestnut on the upper abdomen. In the winter the two bands on the lower surface and the black band on the fore neck are brown. Eye, dark brown. Length of the wing, 5 in.; of the tarsus, 1.25 in. Egg—Greenish grey, speckled with black; length, 1.35 in. Australia, Tasmania, Lord Howe Island, Norfolk Island, New Zealand, and the Chatham Islands.

The banded dotterel has been suspected of migrating from New Zealand to Tasmania in the autumn, as it is common in Tasmania in the winter and leaves in the spring. The species occurs all

Banded Dotterel.

Dotterel.

(*Buller.*)

through eastern Australia, as well as in the Lord Howe and Norfolk Islands. In New Zealand it breeds in August and September, and is equally abundant in the South Island all the year round, showing no signs of migrating, and, as Mr. Handly reports that it is also common in Marlborough throughout the year, it may be safely affirmed that it does not pass regularly to and fro betweers Tasmania and New Zealand.   Mr. Potts says that it is one of the earlier breeder ;, whether of the tussock-clad plain or on riverbed spits, in the lat er case often selecting a dense patch of raoulia for a nesting place. Sometimes the mottled eggs may be discovered on the bare sand; just screened, perhaps, by an uprooted tussock or other waif that has been washed down stream and stranded on the stony flat. He records the fact that it does not appear to be much disturbed by settlement, as it has been known to breed on farms, sometimes in the immediate neighbourhood of the homesteads. The eggs may be found as early as the beginning of August, and the season lasts almost through December, according to locality. It is described as one of the most restless and wariest of birds during the breeding season. On the approach of an intruder, it flies round and round uttering its note of warning; and then, alighting on some rising ground, it steadily keeps watch. During the time it remains on the look out, it has a peculiar habit of jerking its head backwards and forwards uttering its monotonous "twit-twit" at intervals.

Genus Thinornis.

Bill longer than the middle toe with claw. First and second quills nearly equal and longest. Tarsus not longer than the middle toe: hind toe wanting. New Zealand and Auckland Islands.

The Sand Plover.—KUKURUATU.

Thinornis novae-zealandiae.

Above greyish brown, with the forehead, cheeks, throat, and a ring round the nape black. Below white. Bill orange, with a black tip. Eye dark brown. In the young the throat and forehead are white, and the bill is yellow at the base only. Length of the wing, 4.7 in.; of the tarsus, 1 in. Egg variable in colour, cream or buff, with small dark spots and lines; length, 1.3 in. New Zealand and Chatham Islands.

The sand plover was once common in sandy bays from the Great Barrier Island southward to Otago, but it is now very rare. It never went inland. For a nest, it is content to collect a few leaves of grass, bent and twisted into a circular form just large enough to contain the eggs, which are protected by this flimsy structure, as it keeps them together. Mr. Potts adds the following interesting item :—"To the north-by-west of the main Chatham Island lies a small group of rocky islets known as The Sisters, or Rangitutahi. One of these wave-beaten islets, rising

(*Voy. Erebus and Terror.*)

Sand   Plover.

to about 150 feet above the sea, and having an area of only about five acres, affords a nesting place to the sand plover. This very exposed and sheltered site is shared only by the huge albatross and the nelly, which there rest awhile from almost ceaseless wanderings over the surrounding ocean. Exposed to gales that sweep over a vast unbroken expanse of sea and break against this little speck of rock, the only screen that may shelter the sand plover is the tussock of wiry grass or saw-edged carex, for no tree is found there to furnish a kindly shelter." The eggs are three in number. These birds breed in the south, and pass the winter in the North Island.

The Auckland Island Sand Plover.

Thinornis rossi.

Like the last species, but darker in colour, and the feathers on the sides of the body with brown bases. Length of the wing, 4.8 in.; of the tarsus, 1 in. Auckland Islands.

(Voy. Erebus and Terror.)
Auckland Island Sand Plover.

Genus Anarhynchus.

Bill long, slender, bent to the right. First and second quills equal and longest. Tarsus longer than the middle toe; hind toe wanting. New Zealand only.

The Wry-Bill.—Ngutupare.

Anarhynchus frontalis.

Above cinereous grey; below white, with a black band on the breast. Forehead white. Quills brown, with white shafts. Eye dark brown. Length of the wing, 4.75 in.; of the tarsus, 1 in. Egg—Greenish grey, minutely speckled all over with dark brown; length, 1.4 in. Both Islands.

The wry-bill breeds in the South Island, and goes to the North Island in the winter. It is supposed that, in the breeding season, the bird is less wary perhaps than any other species of the family. It displays remarkable instinct in selecting the ground for depositing its eggs. They are laid in a slight depression amongst the pebbles of a river-bed, without any addition of vegetable material, and their grey tint harmonises with the general colour of the shingle. The breeding season extends

Wry-Bill

(*Buller*.)

through spring and early summer. The female usually lays three eggs. Many years ago, Mr. Potts, it might be said, lived amongst them at breeding-time, and, when sleighing drift-wood, he had plenty of opportunities for observing their quaint ways. He says that the parent birds are seldom seen far apart. If disturbed, they trot off at a fast pace, partially opening their wings, which gives a broad flattened appearance across the back. If distressed by too close an approach to the eggs, they utter a low purring sound, carrying the head low, the bill pointing downwards, and just clearing the ground. The young endeavour to escape observation by dropping down close to stones, and they may be readily passed over, as their colour matches that of the pebbles round about them. When frightened, they utter a shrill piping cry, and, if closely pressed, take to the water, for they can swim well. These breeding places, however, it should be stated, are now almost deserted.

The object of the bill being bent to the right has been much discussed, but no satisfactory reason has yet been given for it. One suggestion is that it is used as a lever for overturning stones to get at the insects underneath them; but it must be remembered that a very large part of the bird's life is spent on sand and mud flats. It is only in the breeding season that it feeds among stones.

Genus Himantopus.

Bill much longer than the head, slender, straight. Tarsi very long and slender; toes united at the base by a small membrane; hind toe wanting. Almost cosmopolitan.

Key to the Species.

1. Under surface of body black.	H. melas.
Under surface of body white.	2
2. A white collar round the hind neck.	H. leucocephalus.
No white collar round the neck.	H. picatus.

The White-headed Stilt.—Tuturi-pourewa.

Himantopus leucocephalus.

Above, glossy black; head, under surface, a collar round the back of the neck, and upper tail coverts, white. Tail white, washed with brown on the centre feathers and on the outer webs of the outer ones. Bill black. Legs and feet light red. Eye red. Length of the wing, 9.4 inches; of the tarsus, 4.5 inches. Egg—Yellowish brown, spotted all over with brownish black; length, 1.5 inches. In young birds the back of the neck is grey. H. albicollis of Buller appears to be a variety of this species with the neck all white. Molucca Islands to Australia and New Zealand.

Mr. Potts says that this bird makes a slight nest of grass, often by the edges of lagoons, five or six nests generally being found

together. It lays five eggs, or, rarely, six, and the nesting season
commences in October. He once saw a fight between the old birds
and their last year's young ones, for the possession of the old

White-headed Stilt.

(*Buller*.)

Avocet.   Black Stilt.

nest. "At length the old birds were victorious, after two or
three days' contention, during which time the struggle was
carried on by cries, menacing gestures, and the fluttering of
wings, the birds often leaping over each other. From the long

legs of the birds, their postures and actions in this quarrel seemed very grotesque. The younger and defeated pair retired to about half a chain distant, where they successfully reared a family; this, too, in sight of the passengers on a roadway.''

In the New Zealand bird the legs are not so long as in specimens from Australia, and, perhaps, it ought to be united to the next species.

The Pied Stilt.—POAKA.

Himantopus picatus.

Like the last, but without the white collar on the hind neck. Tail feathers edged more or less broadly with black near the tips. Eye red. Length of the wing, 9.5 inches; of the tarsus, 4 inches; of the egg, 1.8 inch. *H. spicatus* of Potts appears to be a variety with the fore-neck and tail black. New Zealand only.

The pied stilt usually commences to breed in October. Its habits are like those of the last species. Both of them seem to perform some sort of inland migration, as they are found near Napier in the summer but not in the winter, and they leave Canterbury in the middle of May, returning early in August.

The Black Stilt.—KAKI.

Himantopus melas.

Entirely black. Legs and feet red. Eye red. Length of the wing, 9.8 inches; of the tarsus, 3.55 inches. The young are more or less blotched with white, with a good deal of white on the base of the tail, and the throat and breast are white. Egg—Yellowish brown, spotted all over ..ith blackish brown; length, 1.8 inch. New Zealand only.

The black stilt breeds early in the season, on sandy river-beds. It is stated that the labour of nidification is very trifling. Sometimes a nest is made of grass, but, more frequently, a slight depression in a sandy spit answers all requirements as a nesting place. The nest is never very far from the water. The female lays three or four eggs, and the young can run almost as soon as they are hatched. When disturbed, they conceal themselves behind stones, or some other shelter, in the most artful manner. The parent birds exhibit the utmost assiduity in attempting to lead intruders from their eggs or young, and their numerous cunning devices are carried on with surprising cleverness and perseverance. When walking or running over the flats, or wading in shallow water, the stilts are graceful birds; but their flight, with their long legs stretched out behind them, is clumsy. On the wing they constantly utter a sharp cry, something like the bark of a small dog.

Genus Recurvirostra.

Bill long and slender, curved upwards. Tarsi much longer than the middle toe; the toes united by an indented web; the hind toe very short. Cosmopolitan.

The Avocet.

Recurvirostra novae-hollandiae.

Head and upper part of neck chestnut. Middle of the wings, quills, and shoulders, black; remainder white. Tail pale ash. Bill black. Legs, blue. Eye, red. Length of the wing, 9 inches; of the tarsus, 3.5 inches. Australia and New Zealand.

At one time, our avocet was seen fairly frequently in some localities, notably in the swampy lands on the north bank of the Waikerikeri, by the Waimate lagoon, near the mouth of the Rakaia, and on the lagoons near Rockwood, in the Malvern district. Now, however, it is very rare. There are generally four eggs in a nest; they are of a pale yellowish brown colour, and are marked with spots of umber brown and black, with small greyish marks interspersed; their length is 1.8 in. The webbed feet of these birds enable them to walk over very soft mud.

Family Scolopacidae.

Bill long, slender, grooved to the tip. Wings long and pointed. Hind toe short. Eye dark brown in all.

Key to the Genera.

1. Toes with a small web at the base.   Limosa.
   Toes without any web.   2
2. Bill the same length as the tarsus.   3
   Bill longer than the tarsus.   4
3. Tarsus equal to middle toe and claw.   Limonites.
   Tarsus longer than middle toe and claw.   Heteropygia.
4. Tarsus longer than the middle toe.   Tringa.
   Tarsus shorter than the middle toe.   Gallinago.

Genus Limosa.

Bill very long, inclined slightly upwards. First quill the longest. Tarsus longer than the middle toe; outer toe united to the middle as far as the first joint by a membrane; hind toe rather long. Almost cosmopolitan.

The Godwit.—KUAKA.

Limosa novae-zealandiae.

Above greyish brown. Over the tail white, barred with brown. Tail white, barred with greyish brown. Throat brownish white, streaked with darker. Breast rufous in spring, white in autumn. Abdomen white. Shafts of the quills white. Length of wing, 9.5 in.; of the tarsus, 2.5 in. Eastern Siberia, China, Malay Archipelago, to Australia, New Zealand, and Polynesia. Migratory; common on sandy bays and estuaries in the summer.

A great deal has been already written about the godwits, and several descriptions of their remarkable journeys from the North Polar regions to New Zealand have been published. They breed in Eastern Siberia from June to the end of July, and then they leave, passing through country after country, until they reach these southern shores. In September, and again in April, they are found in China, some of them passing the winter in the island of Formosa. Others, in August or September, arrive from the north in Australia, Fiji, New Caledonia, and the New Hebrides, and depart again for the north early in May. Stragglers go to Samoa and Tonga. Many of them arrive in New Zealand in October, November, and December. They spread as far south as Stewart Island, and leave at the end of April or the beginning of May. Visits are also made to the Chatham Islands. The godwits, however, have not been known to breed either in

Australia or New Zealand. They reach this dominion in small parties, which evade observation; but they leave the North Cape and other northern districts in large flocks, which have been

Godwit.

(*Meyer.*)

seen to depart by several observers. A full description of the flight, which is one of the most notable things in natural history, is reproduced in *Nature in New Zealand.**

There is plenty of evidence in regard to the godwits' migration. If any more is needed, it may be found in the change of plumage.

---

* Pages 50-53 (Whitcombe and Tombs).

The godwit is one of those birds that have different plumages in summer and winter. In the Siberian summer, during breeding time, they have their summer plumage; but in New Zealand, although it is summer with us, they are nearly always in their winter plumage. Birds have been noticed here in summer plumage; but it is probable that they remained behind when the great exodus took place, or they may have been birds that moulted early. It is not uncommon for large flocks to remain in New Zealand all the year round. They probably miss one season of migration and wait until the next season comes round.

Sandpiper.　　*(Meyer.)*

Genus Limonites.

Bill about equal to the tarsus, which is about equal to the middle toe and claw.

The Red-necked Sandpiper.

Limonites ruficollis.

Summer plumage—Above chestnut brown, mottled with darker. Sides of the face, neck, and breast rufous; the abdomen white. Winter plumage—Above ashy-brown, mottled with darker. Below white, the upper breast brownish. Length of the wing, 3.8 in.; of the tarsus, 0.75 in. Migratory, but appears to breed sometimes in New Zealand.

Genus *Heteropygia*.

Bill about equal to the tarsus, which is longer than the middle toe and claw.

The Sandpiper.

Heteropygia acuminata.

Summer plumage—Above, sandy rufous streaked with brown. Sides of the face white, with dusky streaks. Below white, the upper breast tinged with sandy rufous, and spotted with black. Winter plumage— Brown instead of rufous. Length of the wing, 5.4 in.; of the tarsus, 1.2 in. Migratory, breeding in the Northern Hemisphere.

The sandpiper is another notable migrant. It breeds in Siberia and Alaska, and, from the former country, passes through Japan and China to the Malay Archipelago, where it spreads into Australia, New Caledonia, and New Zealand. In Australia it is distinctly migratory, while in Tonga it is only a straggler, as it does not visit the islands every year. In New Zealand it is doubtful under which heading it should be placed, for our information about its habits is too scanty, and we do not at present know whether it is or is not an annual visitor to us. It is a fearless little bird, frequenting marshy places, and has a snipe-like flight.

Genus *Tringa*.

Bill longer than the tarsus, which is longer than the middle toe. Inner secondaries considerably shorter than the primaries. Tail square.

The Knot.—HUAHOU.

Tringa canutus.

Summer plumage—Above ashy grey, the upper tail coverts white. Below and sides of the face chestnut. Winter plumage—White below. Length of the wing, 6 in.; of the tarsus, 1.2 in. Migratory, breeding in the Northern Hemisphere.

P

The knot is yet another northern bird which, after breeding in Siberia, travels southwards across the Equator.   Its summer plumage is very different from its winter plumage, which it assumes in September, and retains until May.   In Canterbury, it appears in November, and leaves about April, thus remaining all through our summer.   Generally the birds are in their winter plumage here, but there are two specimens in the Canterbury

Knot: breeding plumage.   (*Meyer.*)

Museum in summer plumage.   One of these was shot on April 2nd, and the other in November, 1899, the latter being in company with others in which the summer plumage is just beginning to show.   They were shot at Lake Ellesmere.   Mr. John Gould mentions a bird from Queensland, shot on September 2nd, 1861, as changing into summer plumage.   It seems, therefore, that some birds have changed the seasons for moulting, and put on their breeding-dress in our summer, and it is probable that these birds breed in New Zealand, although they are not known to remain here through the winter.

Genus *Gallinago*.

Bill long. First and second quills equal, inner secondaries equal in length to the primaries. Tarsus shorter than the middle toe and claw. Two notches in the posterior margin of the sternum. Nearly cosmopolitan.

Key to the Species.

1.	Smaller, wings 3.8.	G. pusilla.
	Larger, wings 4.25.	2
2.	Paler in colour.	G. aucklandica.
	Darker in colour.	G. huegeli.

Auckland Island Snipe.　　*(Voy. Erebus and Terror.)*

The Auckland Island Snipe.

Gallinago aucklandica.

Above, fulvous marked with black. Below, fulvous white, with brown markings on the breast. A brown stripe from the nostrils under the eye to the back of the head, and another in an oblique direction on each cheek. Length of the wing, 4.2 in.; of the tarsus, 1.1 in. Auckland Islands and Antipodes Island.

The Auckland Island snipe are tame, and by no means plentiful. They fly badly, and only for a short distance, rarely more than twenty yards, yet they attempt, on rising, the same zig-zag dashes which make their European congeners so difficult to shoot.   After alighting, they generally run some distance.

The Snares Snipe.

Gallinago huegeli.

Like the last, but darker, and the under surface regularly barred with blackish on the breast and abdomen.   Length of the wing, 4.3 in.; of the tarsus, 0.9 in.   Snares Island.

The Chatham Island Snipe.

Gallinago pusilla.

Above rufous brown, spotted with black and fulvous.   Below brownish white, spotted on the breast with brown.   A brownish white longitudinal line on the top of the head, and a brown line from the nostrils to the eye.   Length of the wing, 3.8 in.; of the tarsus, 0.8 in. Egg—Pinkish buff, with reddish brown spots, closer at the thick end of the egg, where they sometimes form a ring; length, 1.5 in.   Chatham Islands.

Nothing is known about the habits of this bird.

ORDER GAVIAE.

Bill simple, long, the nostrils lateral.   Wings long, primaries ten visible, the fifth secondary wanting.   Tail with twelve feathers.   Front toes entirely connected by webs; the hind toe short and elevated.

Key to the Families.

1. Bill hooked at the tip.   Stercorariidæ.
   Bill not hooked at the tip.   2
2. Tail forked or graduated.   Sternidæ.
   Tail square at the end.   Laridæ.

Family Stercorariidae.

Bill with skin at the base, the tip of the upper mandible hooked. Toes with strong, sharp, hooked claws.

Key to the Genera.
The two middle tail feathers projecting.   Stercorarius.
Two middle tail feathers not longer.   Megalestris.

Genus Stercorarius.

Size smaller, form slender. Tarsus distinctly shorter than the middle toe with its claw. Central tail feathers projecting three inches or more in adults.

Skua Gull.   *(Meyer.)*

The Skua Gull.

Stercorarius crepidatus.

Back and wing-coverts brownish cinereous; top of the head brown; neck and breast white; abdomen dusky. Quills and tail black; two long narrow plumes from the tail. Eye chestnut brown. Length of the wing 13 in.; of the tarsus, 1.75 in. Breeds in the Arctic regions, and passes south in the winter. It has not been recorded from the coasts of the South Island.

Genus Megalestris.

Size larger, form robust. Tarsus rather shorter than the middle toe with its claw. Tail short, the middle feathers only projecting for half-an-inch.

The Sea Hawk.—HAKOAKOA.

Megalestris antarctica.

Brown, finely streaked with pale yellow on the back of the neck.
Basal half of the quills and their shafts white. Eye brown. The pale
yellow streaks on the back of the neck are absent in the young birds.
Length of the wing, 17 in.; of the tarsus, 2.8 in. Egg—Olive brown,
with large purplish grey and brown spots. Length 3 in. The southern
islands of New Zealand and the Chatham Islands.

Sea Hawk.   (*Cat. Brit. Mus.*)

M. antarctica is found all through the Southern Ocean, and in
South America. Here *M. chilensis* connects it with *M.
catarrhactes* of the Northern Hemisphere; so that there can be
very little doubt as to *M. antarctica* having spread from South
America.

Although this bird is often seen at sea, out of sight of land, it
rarely settles on the water, and, when it does so, it holds up its
wings perpendicularly, as if it was afraid of wetting them. It
does not skim over the water as the petrels do, but flies low,
with a heavy slow flapping of its roundish wings, and is, therefore, easily recognised. Its natural food is young and sick birds,
which it kills without remorse, although it is no fighter, but a
great coward.

In the Auckland and Campbell Islands it builds in flat grassy places, not far from the sea, and its small nest is made of grass. The eggs are laid in November or December.

––––––––

Family Sternidae.

Bill straight, rather slender. Both mandibles of about equal length. Tail forked or graduated.

Caspian Tern.

(*Meyer.*)

Key to the Genera.

1.	Tail forked.	2
	Tail graduated.	3
2.	Tail very short.	Hydroprogne.
	Tail more than half the length of wing	Sterna.
3.	Middle toe longer than the bill.	Procelsterna.
	Middle toe shorter than the bill.	4
4.	Bill long and slender.	Micranous.
	Bill stout at the base.	Gygis.

Genus Hydroprogne.

Tail very short, less than one-third of the wing, the outer feathers the longest. Bill very stout and deep. Tarsus less than middle toe and claw.

The Caspian Tern.—TARA-NUI.

Hydroprogne caspia.

White with the back and wings ash-grey. Top of the head black in the spring, but white spotted with black during the rest of the year. Bill red; legs black; eye dusky. Length of the wing, 16 in.; of the tarsus, 1.65 in. Egg—Greyish white, sparingly spotted with dark grey and brown; length 2.6 in. Found over almost the whole world, except Polynesia.

Of this fine bird, Mr. Potts says:—"It is content with merely a hollow scraped in the sand, just large enough to contain the eggs. The breeding season extends from November to January. When the Caspian terns are disturbed at breeding-time, they ascend to a great height, and hover around the intruder, uttering loud screams." The bird's favourite breeding places are spits of land running out from the shore. It is a solitary bird.

Genus Sterna.

Tail very long, more than three-fourths of the wing, the outer feather longest. Bill rather stout. Tarsus less than the middle toe and claw.

New Zealand properly possesses only five kinds of terns, as well as four tropical species which occur at the Kermadec Islands. And of these five, only one species, *S. albistriata,* is peculiar to New Zealand. This species is allied to *S. vittata,* which breeds in the southern islands of New Zealand, as well as in Kerguelen Island and other islands of the South Atlantic. This species, again, is related to *S. macrura* of the Northern Hemisphere, which penetrates southwards to Peru and Brazil in the winter, so that the ancestors of *S. albistriata* probably came to us from South America. *S. frontalis,* which is also found in Australia, seems to be related to *S. bergii* of India and Australia, so that it probably came to us from South Asia. *S. nereis,* which is also an Australian bird, is related to *S. sinensis* of India and the Malay Archipelago. The Caspian tern is widely spread over the world, but is not found in South America, so that we must add it also to our list as coming from South Asia.

The Black-fronted Tern.—TARA.

Sterna albistriata.

Ash-grey, white over the tail. Top of the head and back of the neck black, margined with white. Tail ash-grey; bill orange; legs red; eye black. Length of the wing, 9.5 in.; of the tarsus, 0.7 in. In the young the head is white mottled with black, and the bill and feet are yellow. Egg—Yellowish grey, spotted with grey and brown; length, 1.57 in. Formerly very abundant, but getting rare now in the south. It is a most useful bird, and the eggs and nesting places should be protected.

This beautiful little bird was once seen in large numbers hovering over the newly-ploughed fields in search of larvæ of various insects, and small lizards; but it is rarely seen now. The cause of its decline in numbers is a mystery. A curious habit of this bird has been recorded. A large flock will rest motionless on the ground, with their delicate bluish-grey wings extended vertically, and they will maintain this singular position for some time. The black-fronted tern deposits its eggs on the bare ground, without any attempt at making a nest. It is very clamorous at breeding season, and, if an intruder approaches the ground, he is assailed with swift dartings, accompanied by noisy, harsh, and grating screams.

The Swallow-tailed Tern.

Sterna vittata.

Pale grey; top of the head black. Tail pure white, very deeply forked. Bill and feet cherry-red. Eye black. Length of the wing, 10.5 in.; of the tarsus, 0.7 in. In the winter the top of the head is mottled with grey and black. In the young the bill and feet are reddish black. Campbell, Auckland, and Bounty Islands. Found also at Kerguelen, and the islands of the South Atlantic.

The swallow-tailed terns are noted for their skill in diving, and are characterised by great courage. Dr. Kidder has given a

very good description of these birds as he found them in
Kerguelen Island. He says:—"They dive readily from a
considerable height in the air, rarely missing their mark, a good-
sized crustacean, which seems to constitute their sole diet. During
the pairing season, October, they remind one forcibly of the
common sparrow, curveting round one another, with wings half
spread and constantly chattering. They are bold, showing
scarcely any fear of man. They nest on rather high and broken
ground, usually under the lee of a tuft of grass, and with little
or no preparation. Sometimes a few dried stalks are placed
together in the bottom of a barely perceptible cavity; oftener a
tuft of dead azorella leaves, found ready to hand, serves their
turn. The nests are built not far from the sea, usually on the
slope of a hill-side, where drainage is good, and generally there
are a good many near together. Upon the approach of man, dog,
or sea hawk, a warning scream is sounded, and the whole colony
at once fly up and make common cause against the intruder. The
sea hawk is actually afraid of them, and it is a steady-nerved
man who will not dodge the vicious swoops made from time to
time at his head. So near do they come on these occasions, that
most of my specimens were knocked down with stones when
flying." The eggs are olive-buff, spotted and streaked with
blackish-brown and grey; the length is about 1.8 in.

The White-fronted Tern.—TARA.

Sterna frontalis.

White, with the back and wings pale grey. The top of the head and
back of the neck black. A white line over the bill. Bill black; legs and
feet reddish brown; eye black. Length of the wing, 10.5 in.; of the
tarsus 0.75 in. Egg—Yellowish brown, blotched with grey and black;
length, 1.75 in. New Zealand, Chatham Islands.

This is essentially a sea-coast tern, as it never goes far inland.
It is stated that the liveliness of its movements on the wing,
especially the rapidity with which it drops from a great height
to secure its finny prey, frequently renders it an object of remark
to the dwellers on the sea-shore. It deposits its eggs on the bare
rock, without the slightest protection, at about five or six feet

from the level of high tide, within easy reach of the showers of spray. It also breeds on the shingle-banks near the mouths of rivers.

(*Voy. Erebus and Terror.*)
White-fronted Tern.

The Sooty Tern.

Sterna fuliginosa.

Above, sooty black. Forehead, below, and outer tail feathers, white. Bill and feet reddish black. Length of the wing, 11.75 inches; of the tarsus, 0.9 inch. Egg—Pinkish white, spotted with dark reddish brown and grey; length, 2 inches. Kermadec Islands, tropical and sub-tropical seas.

It lays a single egg on the bare ground, generally in scrub.

The Little Tern.—Tara-iti.

Sterna nereis.

Pale grey above, white below. Top of the head black. Tail white. Bill and feet yellow. Length of the wing, 7.25 inches; of the tarsus, 0.6 inch. In the young the top of the head is white, mottled with black. Egg—White, spotted with grey; length, 1.35 inch. New Zealand and Australia.

It has been recorded that this bird visits the river-bed of the Rakaia, in Canterbury, during the breeding season, not far below the gorge of the river, and that it lays two eggs on the bare ground.

Genus Procelsterna.

Tail graduated, feathers pointed, the outer pair shorter than the next. Foot very long, the middle toe and claw longer than the bill. Second quill the longest. Pacific Ocean.

The Grey Noddy.

Procelsterna cinerea.

Grey, the back darker; the secondaries bordered with white. Bill black. Legs and toes brown; the webs yellow. Eye blackish blue. Length of the wing, 8.5 inches; of the tarsus, 1 inch. Egg—Pinkish cream, sparingly dotted with brown; length 1.7 inch. Kermadec Islands and Australia.

Genus Micranous.

Tail graduated. Bill long and slender. Third quill the longest. Tropical and sub-tropical seas.

The White-capped Noddy.

Micranous leucocapillus.

Black, the top of the head grey. Length of the wing, 8.9 inches; of the tarsus, 0.9 inch. Egg—White or pale pink, blotched principally at the larger end with reddish brown; length, 1.75 inch. Kermadec Islands. Pacific and Atlantic Oceans.

Genus Gygis.

Tail graduated. Bill stout at the base, pointed. Toes slender,
the middle one very long; the webs indented. Inter-tropical seas.

The White Tern.

Gygis candida.

Pure white, with a narrow black rim round the eyes. Bill black.
Legs and feet brown. Eye blue. Length of the wing, 9.5 inches; of the
tarsus, 0.6 inch. The eggs are pale cream or buff, with scrolls, lines, and
specks thickly spread over the egg; length, 1.6 inch; only one is contained
in a nest. Kermadec Islands. Widely distributed.

Family Laridae.

Bill with the upper mandible longer, and bent down over the
tip of the inferior one. Tail usually square at the end.

Genus Larus.

Tail square, or nearly so. Bill more than twice as long as it is
deep. The nostrils linear. Hind toe well developed. Nearly
cosmopolitan, except Polynesia.

The distribution of Larus is very peculiar, as it is not found
in the Malay Archipelago, New Guinea, or the whole of Polynesia;
so that the gulls of Australia and New Caledonia are isolated
from their fellows. Our little mackerel gull, *L. scopulinus*, and
the Australian *L. novae-hollandiae*, are allied to *L. hartlaubi*, of
South Africa and Madagascar; and this again is connected with
L. gelastes, which extends from the Mediterranean to the mouth
of the Indus. It is therefore probable that the ancestors of our
birds came from South Africa. Whether *L. scopulinus* is
descended form *L. novae-hollandiae*, or *vice versa*, or whether
both are branches from *L. hartlaubi*, there is no evidence to
determine.

Our black-backed gull, *L. dominicanus*, is found in South
Africa, as well as in South America and Kerguelen Island, but
not in Tasmania or Australia. It is related to *L. marinus*, of the

Northern Hemisphere, which wanders as far south as Florida; so that probably *L. dominicanus* spread from South America.

At first, it seems singular that our gulls should be different from those of Tasmania and Australia. But we must remember that gulls find their natural food on the seashore, and do not go much to sea except when they are following ships; and even then, it is only the skuas which go out of sight of land. Terns, on the contrary, live entirely on fish or crustaceans, which they take out of the water, and, consequently, they are more oceanic and more widely spread than the gulls.

Black-backed Gull.   (*Gray's Genera of Birds.*)

Key to the Species.

1. Back sooty black.	L. dominicanus
Back pearl-grey.	2
2. Quills white on both sides of the shafts.	L. bulleri.
Inner web of first quill mostly black.	L. scopulinus.

The Black-backed Gull.—KARORO.

Larus dominicanus.

White with black back and wings, the secondaries tipped with white. Eye, white. Young, brown, getting mottled with white as they grow

older. Length of the wing, 16.5 inches; of the tarsus, 2.5 inches. Egg—
Yellowish grey, blotched with grey and dark brown; length 2.85 inches.
Found in New Zealand, the Chatham Islands, and all the southern islands
of New Zealand.

This gull breeds on the sea-shore, and on the sandy spits in the
river-beds. It has a rough-looking nest, which is large, and is
usually made of grass, but sometimes of small tussocks pulled up
by the roots. Mr. Potts states that the parent birds defend their
nest with great spirit. A pair will drive away and chase a
harrier. In the West Coast Sounds the black-backed gull is less
gregarious than on the eastern shores. It feeds principally on
dead animals of all kinds thrown upon the beach; but it is not
averse to live animals when it can get them. It has discovered a
very ingenious way of getting at the soft morsel of the animal of
the pipi-shell (*Mesodesma*). It picks one up in its bill, flies to a
height of fifty feet or so, and then drops the shell on the hard
sand. If it is not broken it picks up the shell and drops it again.
The bird's powers of digestion are great. In captivity it has been
known to swallow mutton-cutlet bones whole, and gradually digest
them. It is a powerful flier, and often soars in circles high up in
the air, like an eagle. It is thought that three or four years elapse
before the black-backed gull gets its full plumage.

The Red-billed Gull.—AKIAKI.

Larus scopulinus.

White, with pearl-grey back and wings. The first and second quills
black with a large spot of white near the tips, most of the others white,
with a black band near the tips. All tipped with white. Bill and feet
red. Eye silvery white. Young, mottled with brown on the back and
wings; tail with a narrow sub-terminal band of brown. Bill brown.
Feet pale red. Length of the wing, 11 inches; of the tarsus, 1.75 inch.
Egg—Yellowish grey, blotched with grey and dark brown; length,
2.1 inches. New Zealand, Chatham Islands, Auckland and Campbell
Islands.

This bird frequents the sea coast, and does not often come
inland, but it is abundant in the harbours. The Auckland
Islands bird, it is stated, has a thicker bill than the bird from New
Zealand. Its flight is light and buoyant, but it rarely goes from

the coast. The constant clamour of the gulls, as they fly, is a striking contrast to the uniform silence of the petrels and albatrosses. Sir Walter Buller says that this bird plunders the oyster catcher in a very systematic manner. "Nature has furnished the last-named bird with a long bill, with which it is enabled to forage in the soft sand for blue crabs and other small crustaceans. The red-billed gull is aware of this, and cultivates the society of his long-billed neighbour to some advantage: he

Red-billed Gull.

(*Meyer.*)

dogs his steps very perseveringly, walking and flying after him, and then quietly standing by till something is captured, when he raises his wings and makes a dash for it. The oyster catcher may succeed in flying off with his prey; but the plunderer, being swifter on the wing, pursues, overtakes, and compels a surrender. The gentleman of the long bill looks gravely on while his crab is being devoured; and having seen the last of it, he gives a stifled whistle, and trots off in search of another, his eager attendant following suit."

Red-billed gulls are very plentiful on the Little Barrier Island, although they do not nest there. They arrive in large flocks and frequent the boulder banks, hundreds of them fluttering down at the same time on the stones and boulders close to the water's edge. Sometimes they stand for half an hour or so, all with their faces to the wind, each on its own little boulder. Suddenly there is a peculiar grating cry, like "kroo kroo!" It is taken up by the whole flock, and then, with a flutter of their wings, they rise about 20 feet from the boulders, like a great white curtain. They flutter about in the air for a few seconds and then settle down on to the boulders again, facing the wind. When they go through this movement, they afford a very pretty sight. They have white breasts, pearly-grey backs, black marked wings, and crimson bills, legs and feet. In their flight, they display much grace. They may often be seen watching for shoals of fish to come in from the sea. Some occasionally wade out into the ripples and allow the water to wash past their legs, like children playing on the sea-shore. As soon as an old bird catches a fish, the young ones set up a noisy "krooing," and open their mouths remarkably wide. The fish is dropped into the open mouth, and is greedily gobbled by the gluttonous young bird. The fish has hardly been swallowed, however, before the "krooing" begins again, and the parents, in a state of great anxiety, start out upon another fishing expedition.

The Black-billed Gull.

Larus bulleri.

Like the last species, but the first three quills are white, margined with black. All tipped with white. Bill black. Legs and feet red. Eye white. Young, mottled with brown on the back and wings. The first and second quills black, with an elongated white spot. Bill black. Legs and feet dark brown. Length of the wing, 12 inches; of the tarsus, 1.7 inch. Egg—Buff or grey, blotched with dark brown; length, 2.2 inches. New Zealand only. Frequents the rivers and lakes, and is rarely seen on the sea coast, although it occasionally frequents the harbours.

According to Mr. Travers, these birds pursue and capture various species of moths, which occur in large numbers among the tussock grass and sedges.

Order Tubinares.

Bill hooked at the tip, the nostrils in tubes. Wing pointed.
Hind toe reduced to a claw. Egg generally white. Eye dark
hazel.

Black-billed Gulls.

(*Buller.*)

Key to the Families.

1. Nostrils on each side of the bill. Diomedeidæ.
   Nostrils on the upper surface of the bill. 2
2. Nostrils distinct. Pelaconoididæ.
   Nostrils united, or nearly so. 3
3. First quill the longest. Puffinidæ.
   Second quill the longest. Procellariidæ.

Family Procellariidae.

Small; the nasal tube long and elevated. In all New Zealand storm petrels the legs and feet are very long and slender, with the outer and middle toes nearly equal. Only ten secondary feathers in the wing.

(Meyer.)

Wilson's Storm Petrel.

Key to the New Zealand Storm Petrels.

1. Upper tail coverts white.	2
Upper tail coverts grey.	3
2. Webs of the toes yellow.	O. oceanicus.
Webs of the toes black.	C. melanogaster.
3. Forehead and throat dark.	G. nereis.
Forehead and throat white.	P. marina.

Genus Oceanites.

Tarsi booted; the claws sharp. World wide.

Wilson's Storm Petrel.

Oceanites oceanicus.

Sooty black; the upper tail coverts and outer feathers of the vent white. Bill black. Legs and toes black, the webs yellow on the inner portion. Length of the wing, 6 inches; of the tarsus, 1.8 inch. Egg—White, with numerous pinkish dots, which sometimes form a zone round one end; length, 1.3 inch.

This bird is not very common in New Zealand seas. Mr. Gould states that it is exceedingly active when flying, its wings being kept fully expanded, and that it also makes considerable use of its feet in patting the surface of the water, with its wings extended upwards, and its head inclined downwards, to gather any food that may present itself floating on the water.

Genus Garrodia.

Tarsus scutellate; the claws sharp. Southern Ocean only.

The Grey-backed Storm Petrel.—REOREO.

Garrodia nereis.

Greyish black, darkest on the head; the tail with a black tip. Abdomen white. Bill, legs, and feet black. Length of the wing, 5.25 inches; of the tarsus, 1.25 inch. Egg—White, with pinkish dots, principally at one end; 1.3 inch in length.

Genus Pelagodroma.

Claws flattened and wide. The first primary much shorter than the third. Southern and Atlantic Oceans.

The White-faced Storm Petrel.

Pelagodroma marina.

Upper surface slaty brown; the upper tail coverts grey. Forehead, a line over each eye, and lower surface white. Bill black. Legs and toes black; the webs yellow in centre. Length of the wing, 5.8 inches; of the tarsus, 1.6 inch. Egg—White, with minute pink dots, principally at one end; 1.4 inch in length.

Genus Cymodroma.

Basal joint of the middle toe much flattened, equal to or longer than the remaining joints and claws. Southern Ocean.

The Black-bellied Storm Petrel.

Cymodroma melanogaster.

Sooty black, the upper tail coverts and flanks white. Breast and middle of the abdomen sooty. Bill, legs, and feet black. Length of the wing, 7 in.; of the tarsus, 1.6 in. Egg—White with pink specks; length, 1.35 in.

Of this species Mr. Gould says: ''When viewed from the ship it is at once distinguished from all the other petrels by the broad black mark which passes down the centre of the abdomen, and offers a strong contrast to the snowy whiteness of the flanks. It is a bird of powerful flight, and pats the surface of the rising waves more frequently than any other species; or perhaps the great length of its legs renders this action more conspicuous. It flutters over the glassy surface of the ocean during calms with an easy butterfly-like motion of the wings, and buffeting and breasting, with equal vigour, the crests of the loftiest waves of the storm; at one moment descending into their deep troughs, and at the next arising with the utmost alertness to their highest points, apparently from an impulse communicated as much by striking the surface of the water with its webbed feet as by the action of the wings. Like the other members of the family it feeds on mollusca, crustacea and any kind of fatty matter that may be floating on the surface of the ocean.''

Family Puffinidae.

Key to the Genera.

1. Palate without lamellæ.   2
    Palate with lamellæ on each side.   6
2. Tarsi compressed.   3
    Tarsi rounded.   5
3. Both nostrils visible.   Puffinidæ.
    Nostrils united in a single opening.   4

4. Tail feathers twelve.	Priofinus.
Tail feathers fourteen.	Priocella.
5. Bill more or less yellow.	Majaqueus.
Bill black.	Œstrelata.
6. Large: tail with sixteen feathers.	Ossifraga.
Smaller: tail with 12 or 14 feathers.	7
7. Black and white.	Daption.
Pale grey above, white below.	8
8. Tip of the tail, white.	Halobæna.
Tip of the tail, dark.	Prion.

Genus Puffinus.

Tarsi distinctly compressed, the anterior edge sharp. Nasal tube low; both nostrils visible from above, directed forwards and slightly upwards. Tail feathers twelve. Cosmopolitan.

Key to the Species.

1. Lower surface dark.	2
Lower surface white.	5
2. Tail long and cuneate.	P. chlororhynchus.
Tail moderate, rounded.	3
3. Bill yellow.	P. carneipes.
Bill dark coloured.	4
4. Under wing coverts pale grey.	P. griseus.
Under wing coverts brownish.	P. tenuirostris.
5. Under tail coverts white.	6
Under tail coverts partly grey.	7
6. Wing from flexure 8 in.	P. gavia
Wing from flexure 7½ in.	P. assimilis.
7. Wing from flexure 11 in.	P. bulleri.
Wing from flexure 8 in.	P. obscurus.

The Long-tailed Shearwater.

Puffinus bulleri.

Upper surface dark grey; the under surface white. Outer half of the inner webs of the primaries white. Tail, long and cuneate. Bill, dark horn colour; legs and toes, yellowish. Length of the wing, 11.3 in.; of the tarsus, 2 in. New Zealand only. Very few specimens have as yet been obtained.

The Wedge-tailed Shearwater.

Puffinus chlororhynchus.

Sooty brown. Bill, flesh colour: the tip dark. Legs and toes, flesh colour; the outer, darker. Length of the wing, 11.6 in.; of the tarsus, 1.8 in. Indian and Central Pacific Oceans. In New Zealand found only in the Auckland district. Egg—2.5 in. in length.

Rain Bird.

Long-tailed Shearwater. *(Buller.)*

The Shearwater.—HAKOAKOA.

Puffinus gavia.

Above slaty black; below, white. Bill, dark horn colour. Outer side of the legs and outer toe, black; inner sides and other toes, with the webs, yellow. Length of the wing, 8 in.; of the tarsus, 1.7 in. Egg—

2.3 in. in length. New Zealand and Australian seas. In New Zealand it is more common in the South Island than in the North.

It is recorded by Mr. Reischek that the shearwaters come on shore to clean out their burrows or make new ones in September. He states that they dig with their bill, pushing out the soil with their feet; working all the day, and, after sunset, leaving for the sea. The burrow is about four inches and a half in diameter, and from a foot and a half to three feet long. At the end there is a chamber, about a foot and a half in diameter with leaves on the bottom, on which, in October, the female lays one white egg. She sits during the day, when the male is usually at sea; and he returns soon after sunset and relieves his mate. This process is continued until the young birds are a few days old, when both parents absent themselves during the day, but return after sunset to feed their young with an oily fluid which they disgorge. The young birds are full grown in March, when they leave their breeding places for the ocean.

The Dusky Shearwater.

Puffinus obscurus.

Above, slaty black; below, white. Under tail coverts, dark brown tipped with white. Bill brown. Outside of leg and outer toe, black; inner side and other toes, yellow. Length of the wing, 7.8 in.; of the tarsus, 1.5 in. Tropical and sub-tropical seas of the whole world. Rare in New Zealand.

The Allied Shearwater.

Puffinus assimilis.

Above, slaty black; below, white. Under tail coverts, pure white. Bill black. Legs and toes, blackish; the webs, yellow. Length of the wing, 7.4 in.; of the tarsus, 1.5 in. Egg—1.95 in. in length. Australian and New Zealand seas. In New Zealand it is not found south of Hauraki Gulf.

The allied shearwater commences to breed in October. Its burrow is about four inches in diameter, and from three to four feet long, and generally has two chambers at the end. The egg is laid at the end of October or the beginning of November, and the young birds are full grown in February.

The Pink-footed Shearwater.

Puffinus carneipes.

Dark sooty brown. Bill, legs, and feet, flesh colour. Tail, rather short and rounded at the end. Length of the wing 12.5 in.; of the tarsus, 2.25 in. Egg—2.8 in. in length. Australia and the northern parts of New Zealand.

Allied Shearwater.   (*Meyer.*)

The Tasmanian Mutton Bird.—Hakoakoa.

Puffinus tenuirostris.

Sooty brown. The under wing coverts, brownish. Bill, horn colour. Legs and feet, yellowish. Length of the wing, 10.4 in.; of the tarsus, 2 in. Egg—2.5 to 2.9 in length. Australian and New Zealand seas. Also Japan and Samoa. It is by no means a common bird in New Zealand.

The attention of naturalists has frequently been turned to the habits of these birds, especially to their gregariousness and flights. A recent writer states that on the eastern coastline of Victoria the mutton birds arrive in countless thousands, coming in immense flocks, and reaching their rookeries within a few hours of the same date every year. In the middle of November, in 1902, for instance, there was no sign of a mutton bird on Phillip Island, which faces Bass Strait and the Southern Ocean. But, on the evening of November 24th, they flew in from the sea in large flocks. An incident related by Captain Waller, of the *Westralia*, illustrates the gregariousness of these birds. He reports that on one occasion, while on the passage from New Zealand to Australia, he steamed for thirty miles through flights of mutton birds, which extended for three or four miles on each side of the vessel. Occasionally they settled on the water to feed, and then they covered the surface, and looked like a reef of black rocks. They were on their way to the Victorian coast, to find a nesting-place in their rookeries. The nests are made in the soft earth.

New Zealand Mutton Bird.   (*From a Specimen.*)

The New Zealand Mutton Bird.—Oɪ.

Puffinus griseus.

Sooty brown, the under wing-coverts pale grey, each feather with a dark shaft. Bill horn colour; legs and feet brown. Length of the wing, 12 in.; of the tarsus, 2.4 in. Egg—2.65 in. in length. North Atlantic and North Pacific; also the Straits of Magellan, New Zealand, Chatham and Auckland Islands.

In the southern parts of New Zealand this bird exists in countless numbers.

Genus *Priofinus*.

Nasal-tube well developed, the nostrils united into a single opening directed forwards. Tarsi compressed. Tail feathers twelve.

Brown Petrel.

(After Gould.)

The Brown Petrel.—KUIA.

Priofinus cinereus.

Above cinereous, below white. Bill yellow on the sides, the nasal-tube and upper margin black; legs and feet flesh colour. Length of the wing, 13 in.; of the tarsus, 2.3 in. Southern Ocean.

This bird combines the appearance of an Œstrelata with some of the habits of Puffinus. Its feathers fit very close, and have a glossy look. Like all other petrels, it flies with its legs stretched straight out behind, and as in this bird they are rather long, they make the tail appear forked. The cry is something like the bleating of a lamb. They are very common at sea from May to August, but retire to breed in September or October.

The brown petrel is by far the best diver of all the sea-going petrels. It seems even to be fond of diving, and often remains under water for several minutes, then it comes up again shaking the water off its feathers like a dog. Sometimes, as it flies past a vessel, it will poise itself for a moment in the air at a height of twenty or twenty-five feet above the sea, and, shutting its wings, take a header into the water. It dives with its wings open, and uses them under water much in the same manner as when flying.

It is sometimes called "Bully" by sailors, and is very plentiful in the South Pacific Ocean, between New Zealand and Cape Horn.

Genus Priocella.

This genus differs from Priofinus in having fourteen tail feathers.

The Silver-grey Petrel.

Priocella glacialoides.

Above pale grey, below white. Bill yellow, the nasal-tube and the tip dark. Legs and feet flesh colour. Length of the wing, 12.6 in.; of the tarsus, 1.8 in. The true home of this bird is the ice-pack of the Antarctic seas, and only stragglers reach New Zealand.

Genus Majaqueus.

Bill long, stout, more or less pale coloured, the hook large; the nasal tubes almost united into a single opening directed forwards. Tarsi rounded on the anterior edge. Tail rounded, composed of twelve feathers. Southern Ocean.

Key to the Species.

Larger, chin white.	M. æquinoctialis.
Smaller, chin black.	M. parkinsoni.

The White-chinned Petrel.

Majaqueus aequinoctialis.

Sooty black, the chin white. Sometimes an irregular white stripe under the eye, and another across the forehead. Bill yellowish horn, varied with black; legs and feet black. Length of the wing, 15 in.; of the tarsus, 2.6 in. Egg—About 3 in. in length. Southern Ocean, south of Lat. 30° South.

This bird breeds in holes at the Auckland Islands. By the sealers it is called stink-pot, or stinker.

The Black Petrel.—TAONUI or TAIKO.

Majaqueus parkinsoni.

Sooty black. Bill with the middle parts bluish white. Legs and feet black. Length of the wing, 13.2 in.; of the tarsus, 2.2 in. Egg—2.85 in. in length. New Zealand seas.

My first acquaintance with the black petrel was on the Little Barrier Island, in February, 1907. During the first night I spent on the sanctuary, I heard this bird's call, which is a strange and unearthly sound. When darkness falls upon the island and all the children of the forest have retired for the night, the black petrel hurries on its flights, returning home across the seas to its subterranean nest on the mountain top, or setting out upon some mysterious voyage through the night. It is as dark as night itself, and is quite invisible, but its cry can be heard far above, and by that means its course through the air can be traced until it reaches the mist-covered mountain-tops. The cry is quite unlike that of any other bird. As the sound comes down through the darkness from a great height, it seems to belong to some ominous, inauspicious creature, whose presence bodes nobody any good. It strikes the ear as the combination of a soft whistle and a deep ''whir,'' coming from the bottom of a husky throat. It is repeated at frequent intervals, and is loud and rasping, and utterly unmusical. I heard this call night after night before I saw the parkinsoni itself. On one occasion, one of these birds fluttered near me as I was going to my tent at night, and I was startled with the noise of its long-sweeping wings, which seemed to come suddenly out of the darkness and rush past with incredible swiftness. Several days afterwards I had an opportunity to meet the parkinsoni at close quarters. It was when I went up to the top of the Herikohu mountain with the members of Mr. Shakespear's family. The mountain rises to a pinnacle peak 2100 feet above the sea. In the soft loose soil on the top of the peak, a member of the party found a parkinsoni's nest. It was a small chamber at the end of a burrow about two feet six inches long. A few leaves had been taken into the nest and placed on the floor. Apart from that, absolutely no attempt

Black Petrel.   (*Cat. Brit. Mus.*)

had been made to create any home comforts for the young. A female bird was sitting on the nest. There is a strict rule on the island that no native birds shall be disturbed or interfered with in any way, but this was relaxed to allow me to handle the parkinsoni, which I was very pleased to do. The bird strongly objected to being touched, and fought furiously with claws, wings, and beak, trying to snap at the fingers that held it. Its plumage was glossy, satiny black; its eyes were large, black, lustrous, and piercing; and it had jet black legs and feet. It was, in fact, entirely black, from the top of its head to the tip of its tail, except its bill, which was bluish white in the middle and a dirty horn yellow at the base. The female lays only one egg at a time. There was an egg in the nest we had discovered, and a chick, which had just thrust its head through the shell, was chirping plaintively. The egg was placed in the nest again, and the struggling bird was set upon it, evidently to her great satisfaction.

As far as is known, these birds are not found out of New Zealand seas. They seem to be more numerous in Hauraki Gulf, Auckland, than in any other part, and they have not been recorded from Canterbury. Large numbers resort to the Little Barrier, and to other islands close by, where they breed. When Mr. Reischek visited the Little Barrier in 1882, he saw several black petrels. One of the nests he inspected was evidently in the same place as the nest I saw, 25 years later. He states that when the birds are not breeding, two of them are often found associated in one hole, but when the nest contains an egg, the female alone is left in charge. In November, he has seen the old birds working together clearing out the hole they had selected and adapting it to their requirements, and afterwards collecting dry leaves and pieces of moss to place in the chamber, or depression at the end of the burrow. The late Mr. T. Kirk has found black petrels on several islands in the Hauraki Gulf. He says that they are always tame, and that they allow themselves to be seized in their burrows without resistance. Mr. Reischek, however, states that he has always found them to be exceedingly fierce, and that is the impression I received from my acquaintance with these birds.

Mr. Reischek states that he has seen them performing feats of expert climbers. With their sharp claws, bills, and wings, they climb trees out of the perpendicular, from which they fly away.

In former years, before the island had been declared a sanctuary, Maoris visited it to collect these birds in April and May, for food supplies. All that kind of thing, however, has been put a stop to as far as the Little Barrier is concerned, and no "mutton birding" is carried on there now on any pretence whatever.*

---

* From a paper on " The Little Barrier Island," by J. Drummond, read before the Philosophical Institute of Canterbury, October 2rd, 1907.

Bill rather short, deep and black; the hook large; nasal opening directed slightly upwards. Tail moderate, rounded, with twelve feathers. Tarsi rounded. Claw of the hind toe small. Temperate and tropical oceans.

Genus Œstrelata.

Key to the Species.

1. Exposed portion of the outer primary dark beneath.	2
Exposed portion of the outer primary white near the base of the inner web.	4
2. Brown below.	Œ. macroptera.
White below.	3
3. Head white.	Œ. lessoni.
Head grey.	Œ. nigripennis.
4. Under wing coverts dark.	Œ. neglecta.
Under wing coverts white.	5
5. Wing over 11 inches.	6
Wing under 9 inches.	7
6. Abdomen white.	Œ. cervicalis.
Abdomen grey.	Œ. inexpectata.
7. Axillaries white.	Œ. cooki.
Axillaries black.	Œ. axillaris.

The Grey-faced Petrel.—Oi.

Œstrelata macroptera.

Dark sooty brown, with a grey face. Bill and feet black. Length of the wing, 12 inches; of the tarsus, 1.65 inch. Egg—2.6 in. in length. Southern Ocean.

The grey-faced petrel breeds in holes in the ground on the islands in Hauraki Gulf and Bay of Plenty, the holes being near the bottom of the cliffs. The eggs are laid in July to September. At sea this bird is solitary and wild, never coming near a vessel; and on the wing it looks like a large swift.

The White-headed Petrel.

Œstrelata lessoni.

Back and tail grey. Wings black. Head and under surface white, with a black line through the eye. Length of the wing, 12 inches; of the tarsus, 1.8 inch. Egg—2.8 inches in length. Southern Ocean.

This bird breeds in holes on the Antipodes Island. Eggs are laid in January. It is solitary and wild, like the grey-faced petrel.

The Black-winged Petrel.

Œstrelata nigrepennis.

Upper surface slaty grey, the crown mottled with black. Forehead and spot over the eye white. Under surface white, the sides of the breast slaty grey. Under surface of the wings white, with a dark brown border. Length of the wing, 8.7 inches; of the tarsus, 1.2 inch. Kermadec Islands.

Grey-faced Petrel.

(After Smith.)

The Black-Capped Petrel.

Œstrelata cervicalis.

Upper surface greyish black. The crown of the head and nape, rusty black. Under surface and front, white. Tarsi and basal part of the feet, yellow; the tips black. Length of the wing, 11.5 inches; of the tarsus, 1.5 inch. Egg—2.5 inches in length. Kermadec Islands.

This petrel arrives about the end of September, and remains until the end of June. It is one of the last petrels to leave the Islands. It is solitary in its habits, and very seldom can two nests be found in the same locality. Its breeding-place is usually near the mountain-top, in some dark gully filled with palms and

R

tree-ferns, and generally its burrow is made at the foot of the latter. It is altogether nocturnal in its habits, and rarely leaves its burrow in the daytime, and therefore it is not seen at sea.

(Cat. Brit. Mus.)

Black-capped Petrel.

The Kermadec Island Mutton Bird.

Œstrelata neglecta.

Very variable in colour. Back, brown. Head and lower surface sometimes white, sometimes brown, or sometimes partly one and partly the other. Under wing coverts and axillaries, brown. Primaries, blackish brown. The bases of the inner webs and the adjoining shafts, white. Bill black. Tarsi and basal portion of toes, yellow; their tips black. Length of the wing, 11.3 inches; of the tarsus, 1.5 inch. Egg—2.55 inches in length. Kermadec Islands.

This species does not form burrows, like the other members of the genus, but breeds in the open. The great variation in plumage is very remarkable; it has not yet been explained. It is certain, however, that the dark birds are not the young of the light ones. Accounts are rather confused, but it seems probable that three varieties can be recognised, with different habits.

These are as follows:—Variety A (*leucophrys*), the lower surface, and often the head, white. Variety B (*neglecta*), the lower surface white, except a band of brown on the breast; head brown. Variety C (*philippi*), the lower surface brown, like the upper. A is the winter mutton bird of the settlers. It breeds principally upon Meyer Island. In regard to the time of the arrival of the old birds Mr. T. F Cheeseman says that he found nearly full

(*Ibis.*)

Kermadec Island Mutton Bird.

grown fledglings in August sitting at the roots of trees. The other two varieties, known as the true mutton bird, arrive in great numbers at the end of August and September, and breed chiefly on Sunday Island. Captain Bollons informs us that members of Variety B commence to breed rather earlier than the others, and that they make their nests inland, on little ridges; while Variety C always breeds on the edge of the sea-cliffs. It seems probable that we have here a very interesting example of the evolution of new species by isolation, due to alterations in the time of breeding of certain individuals.

The Rain Bird.

Œstrelata inexpectata.

Upper surface, dark grey. Abdomen, brownish grey. Throat and under tail coverts, white. Inner webs of the primaries, abruptly white for at least the inner half. Legs and basal portion of the toes yellow; the remainder, black. Length of the wing, 10 inches; of the tarsus, 1.2 inch. New Zealand seas.

The rain bird breeds in holes near the tops of the mountain ranges, far inland, in both islands. It is known as the rain bird in many places, as it is often heard at night calling as it flies to

Cook's Petrel.

(*Voy. Erebus and Terror.*)

the sea; and in New Zealand rain often falls in the night, and the weather is fine in the morning. However, several other petrels do the same thing, and the cry is not restricted to this petrel. (For the figure, see page 247).

Cook's Petrel.—Titi.

Œstrelata cooki.

Above, grey; forehead, cheeks, and below, white; a brown spot through the eye. Under wing coverts, white. Bill slender. Legs and base of toes, yellow; remainder, black. Length of the wing, 9 inches; of the tarsus, 1.2 inches. New Zealand and Japan.

This bird lives on the Chickens Islands, sometimes in the same burrows as tuataras. Mr. Reischek says that the burrows are from four to six inches in diameter, and from four to twelve feet long, each with two chambers, about a foot in diameter, with leaves, moss, or fine grass at the bottom. "After sunset," to use

Chatham Island Petrel.

(*Cat. Brit. Mus.*)

his own words, "they begin to call, like 'ti, ti, ti,' repeated rapidly, which is the signal to assemble for their departure to their ocean haunts, from which they do not return till before sunrise." The eggs are laid early in November, and the young are full grown in March.

The Chatham Island Petrel.

Œstrelata axillaris.

Like the last species, but the bill stouter and the secondary under wing coverts, as well as the axillaries, black. Length of the wing, 8.3 in.; of the tarsus, 1.2 in. Chatham Islands.

Genus Ossifraga.

Very large. The nasal tube very long and stout. Chin feathered throughout. Wings rather short. Tail with 16 feathers. Southern Ocean.

The Nelly.

Ossifraga gigantea.

Brown, the chin and throat paler. Bill, yellow. Legs and feet, brownish black. Length of the wing, 20 in.; of the tarsus, 3.5 in. Egg—4 in. in length.

Nelly.

(*After Gould.*)

This bird breeds in the cliffs of the southern islands late in December, and its nest is obtained only with difficulty. The young are at first covered with long light grey down. When fledged they are dark brown, mottled with white. Albino varieties are common. When a person approaches the nest, the old birds keep a short distance away, while the young ones squirt a horribly smelling oil out of their mouths to a distance of six or eight feet. The bird is very voracious, hovering over the sealers when they are engaged cutting up a seal, and devouring the carcass the moment it is left, a thing that the albatross never does. It has sometimes been seen to chase other birds, but never to kill one. Its flight is much more laboured than that of the albatross. It is a filthy bird, and no one has anything to say in its favour. It has well earned its title of "The Vulture of the Seas." Where quantities of refuse have been deposited, it may often be found gorging itself to repletion. Sailors have asserted that it will attack drowning men, and will begin its horrid repast upon them before they are dead. In Antarctica, nellies seem to be on friendly terms with the penguins, although, when a penguin dies, a nelly that happens to be in the neighbourhood soon gobbles it up. Penguins that were shot by an exploring party in the Antarctic Regions were eaten by a nelly before there was time to row a boat round a piece of rock to pick them off.

Genus Daption.

Bill rather slender, the nasal tube narrower and lower at the base. Tail feathers, twelve. Tarsi, slender. Southern Ocean.

The Cape Pigeon.

Daption capensis.

Head, black; back, white spotted with black; below, white. Bill and feet, black. Length of the wing, 10.5 in.; of the tarsus, 1.8 in. Very common at sea during the winter months, but retires altogether in the summer to breed.

The Cape pigeon evidently breeds on the Snares, Auckland, and Antipodes Islands, as large numbers may be seen there in the summer. These and other small oceanic birds are much more easily caught with a thread than with a hook. The method adopted is as follows :—A small piece of wood, about an inch and a half long, is tied by its middle to a line of white thread or silk. This is thrown over the stern, and is allowed to float out some twenty or thirty yards. The birds, flying under the stern of the ship, strike against the thread and entangle their wings in it. They are then hauled gently on board. If the ship is going very fast, the thread will not be strong enough to hold them; but if it is too thick, they will see it and avoid it.   When caught and brought on board, the birds throw up from their mouths, as soon as touched, a quantity of red, strong-smelling oil, not as a means of offence or defence, but simply from fright.   They cannot rise from the deck, but run along with outstretched wings.   Their cry

(*After Gould.*)

Cape Pigeon.

is like the sound made by drawing a piece of iron across a large toothed comb: "cac, cac, cac-cac, cac," the third being pronounced the quickest.

An interesting feature of the operations of the Scottish National Antarctic Expedition, which was sent out in 1903, and returned in 1904, was the discovery of the egg of the Cape pigeon. Its egg, as far as is known, had never been seen before by human beings.   Like most of the petrels, the Cape pigeon lays a single egg, which is pure white in colour, and is deposited in a nest that consists of a few fragments of stone raked together on a bare ledge of the cliff.

Genus Halobaena.

Small. The nasal tube short. The lamellæ on the sides of the palate rudimentary. First and second primaries sub-equal. Twelve tail feathers. Southern Ocean.

The Blue Petrel.

Halobaena caerulea.

Above, pale ashy blue. The forehead and under surface, white. Tail, grey tipped with white. Bill, black. The edge of the mandible, blue. Tarsi and toes, blue. The webs, flesh colour. Length of the wing, 8.5 in.; of the tarsus, 1.3 in. Egg—2 in. in length. A rare bird in the New Zealand seas.

The Whale Birds.

Genus Prion.

Small. The nasal tube short. The lamellæ on the sides of the palate well developed. First primary the longest. Twelve tail feathers. Southern Ocean.

The whale birds generally fly in flocks, with a zig-zag movement and a sharp motion of the wings, like a snipe, and are rarely seen to sit on the water. They are easily recognised when flying by the dark mark, like W, on their expanded wings. They breed in holes on all the islands round New Zealand, and the egg is about 1.6 in. in length.

Key to the Species.

1. Wing from flexure, 9 to 10 in.	P. vittatus.
Wing from flexure, 8 to 9 in.	2
Wing from flexure, 6 to 7 in.	P. ariel.
2. Upper mandible convex on the margin.	P. banksi.
Upper mandible straight on the margin.	P. desolatus.

Prion vittatus.

Above light grey; over the eye and the lower surface white. Shoulders and tip of the tail brownish black. Bill very wide, edges of the maxilla distinctly convex; lamellæ distinctly visible when the mouth is shut. Bill blue; legs and feet light blue. Chin naked. Length of the wings, 9.5 in.; of the tarsus, 1.3 in.; breadth of the bill at the end of the nasal tube 0.8 in. Egg—1.9 in. in length. Breeds on the Chatham Islands.

Prion banksi.

WHIROIA.

Similar to the last in colour but smaller, and the lamellæ of the maxilla visible near the base only when the bill is closed. Chin half feathered. Length of the wing, 8.5 in.; of the tarsus, 1.3 in.; breadth of the bill at the end of the tube, 0.5 in. Breeds on the Auckland Islands.

Whale Birds.

(After Gould.)

Prion desolatus.

Similar to the last in colour, but with a smaller bill. The sides of the maxilla are nearly straight, and the lamellæ are not visible when the mouth is closed. Chin half feathered. Length of the wing, 8.5 in.; of the tarsus, 1.3 in.; breadth of the bill at the end of the tube, 0.3 to 0.35 in. Egg—1.9 in. in length. Breeds on Antipodes Island.

Prion ariel.

Similar in colour to the others, but with a paler crown. Bill much narrower and more compressed; the sides of the maxilla nearly straight. Chin fully feathered. Length of the wing, 6.8 in.; of the tarsus, 1.2 in. Breeds on the northern part of New Zealand. Much has yet to be learned about the breeding-places of the whale birds.

Family Pelecanoididae.

The Diving Petrel.

Genus Pelecanoides.

Bill shorter than the head; the nasal tubes distinct on the top of the bill, opening vertically upwards. Wings short

The diving petrel does not fly like its congeners, but flutters along, hardly rising above the surface of the water, soon settling again, and generally diving at once. It flies under the water like a penguin. It is gregarious, living in small flocks all round the New Zealand coasts. There are two species.

Pelecanoides urinatrix.

Above brownish black; below, white. Bill, black. Legs and feet pale blue. Length of the wings, 4.7 in.; of the tarsus, 1 in. Egg—1.5 in. in length. Found also at Cape Horn.

Pelecanoides exsul.

Like the last, but the feathers of the sides and middle of the throat with a distinct sub-terminal grey bar; flanks mottled with grey, and under wing-coverts grey. Found also at the Crozette Islands and Kerguelen Land.

Family Diomedeidae.

Bill long, the nasal tubes disjointed, lateral. Wings long and narrow. Hind toe absent. Egg white, often spotted with reddish brown.

Sailors apply the name albatross to the large species with white backs, and distinguish the smaller forms with black backs as mollymawks. The breeding habits in these two groups are very different, the albatrosses choosing grassy flats, the mollymawks rocky cliffs, on which to make their nests. In the Pliocene Period, albatrosses inhabited the North Atlantic Ocean; but at present they are limited to the North Pacific, the coast of Peru, and the Southern Ocean between 30° S. and 60° S. Several are dark coloured when they are young, and get whiter as they grow old;

and this points to the probability of *D. nigripes,* which remains dark all through life, being nearer to the prototype albatross than any other now living.   It seems probable, therefore, that albatrosses originated in the Northern Hemisphere, and passed south through the Pacific Ocean.

No two species of albatross or mollymawk are known to breed in the same locality.   Even when two different kinds are found on the same island, as *D. exulans* and *D. regia,* on Adams Island of the Auckland group, they occupy widely separated sites, and, as will presently be seen, there are no fewer than five months' difference between the first to breed and the last.

As these birds all live on the same food, and have the same simple habits when they are at sea, it cannot be supposed that their distinctive specific characters are due to natural selection, for that which would favour one would favour all.   Nor can it be supposed that they are due to the action of external conditions, because what would affect one would affect all.   Nor, again, can it be supposed that they are recognition-marks, for, when the breeding-time is drawing near, each bird goes separately to its old nest before courtship begins.   The pink feathers on the sides of the neck of *D. chionoptera* may possibly be due to sexual selection, but all the differences between the species do not have such an origin.   The birds appear to mate for life, so that there is very little opportunity for choice.   It cannot, therefore, be that the species of albatrosses were formed by competition on the ocean, and subsequently chose separate breeding-grounds.   It may be believed that isolation preceded the development of their specific characters.

It is not difficult to imagine that those birds to which the breeding impulse came first should retire to their breeding-grounds and there mate; while those in which the impulse was delayed might find their old breeding-grounds fully occupied and would have to choose others.   Thus, owing to physiological isolation, a small number of birds would become physically isolated, and new specific characters might arise and be preserved. This method of physiological isolation has often played an important part in the origin of species without any help from natural selection, not only in birds, but also in insects. It is evident that in an equable climate, where the exact time of breeding was not very important, many variations might be preserved by this means; while in more rigorous climates, where the breeding-season must necessarily be short, this kind of physiological isolation could not occur. And this may account for the greater number of species in tropical countries, especially on islands, as contrasted with the enormous number of individuals belonging to very few species, which is characteristic of temperate regions with continental climates.

Key to the Genera.

1.	Tail pointed at the end.	Phœbetria.
	Tail rounded at the end.	2
2.	Anterior margin of under surface of wing white.	Diomedea.
	Anterior margin of under surface of wing dark.	Thalassarche.

Genus Diomedea.

Large. The back white or brown. Tail short and rounded. Southern Ocean and North Pacific.

Key to the Species.

1.	Upper back with transverse black lines.	D. exulans.
	Upper back without transverse black lines.	2
2.	Wing coverts dark grey and white.	D. regia.
	Wing coverts nearly white.	D. chionoptera.

The Wandering Albatross.—Toroa.

Diomedea exulans.

White. The back transversely pencilled with black. Quills, black. Eyelids, pale purple. Bill, flesh colour (horn colour when dry). Legs and feet flesh colour. Length of the wing, 25.5 in.; of the tarsus, 4.8 in. Egg—About 5 in. in length. Young, chocolate brown, with a white face. Under wings coverts and axillaries, white. Southern Ocean. Breeds on Adams Island, Auckland Group, and at the Antipodes. Its range extends through the Southern Pacific Ocean to within about 600 miles of Cape Horn, after which it appears to be replaced by the snowy albatross (D. chionoptera).

The albatross takes three or four years to attain the adult plumage. The first year is spent in the nest; after leaving it the young bird becomes gradually white, first on the abdomen, then on the back, then on the neck, and last on the top of the head. The Antipodes Island birds may be a distinct species, as they are darker on the back and do not appear to get a white head.

The average breadth across the wings is ten feet, ranging between nine and twelve feet. The bird's food consists chiefly of

(Photo. by Mr. G. Buddle, on Antipodes Island.)
Young of Wandering Albatross.

cuttle-fish. Indeed, it cannot catch fish; for it never pounces suddenly, like a frigate-bird, or a gannet, on anything floating on the water, but always settles first at some little distance off and swims up to its prey. For this reason it can be caught with a hook only when a vessel is going slowly, not more than four or five knots, and when plenty of line can be paid out, so as to give the bird time to look at the bait before he swallows it; for the albatross is a cautious bird, and is frightened if the bait is being towed through the water. The bait must be floated by means of corks; the hook need not be barbed, as it always catches in the curved end of the upper mandible.

The albatross does not fly by night, and its habits are quite diurnal both on sea and on land. It is rarely found north of latitude 30° S. The nest, which is always placed on high grassy tablelands, is shaped like the frustrum of a cone, with a slightly hollowed top, and is made of grass and mud, which the birds obtain by digging a circular ditch, about two yards in diameter, and pushing the earth towards the centre, until it is about eighteen inches high. In this nest the female bird lays one white egg, in the first week in January at the Auckland Islands, and in the middle of January at Antipodes Island. Both sexes sit alternately.

In autumn the old birds leave the breeding-ground and go to sea. When they return, each pair goes at once to its old nest; and, after a little fondling of the young one, which has remained on the nest the whole time, they turn it out and prepare the nest for the next brood. The deserted young one is in good condition and is very lively, and is seen off the nest exercising its wings. When the old birds return and take possession of their nest, the young one

Young Albatross in nest, Antipodes Island.

often remains outside and nibbles at the head of the old bird until the feathers between the beak and the eye are removed, and the skin made quite sore. The young birds do not go far from land until the following year, when they accompany the old ones to sea. It used to be a puzzle how the young birds were fed during the absence of their parents, but when the French Transit of Venus expedition visited Campbell Island in 1874, they placed sentries over some of the nests, when it was found that the old birds visited and fed their young soon after dawn every morning.

The flight of the albatross is truly majestic. With outstretched and motionless wings, he sails over the surface of the sea, now rising high in the air, now, with a bold sweep and wings inclined at a high angle with the horizon, descending until the tip of the lower wing almost touches the crest of the waves as he skims over them. Suddenly he sees something floating on the water, and prepares to alight; but how changed he now is from the noble bird, who, but a moment before, was all grace and symmetry. He raises his wings, his head goes back, his back goes in; down drop two enormous webbed feet, straddled out to their full extent, and then with a hoarse croak, between that of a raven and that of a sheep, he falls souse into the water. There he is at home again, breasting the waves like a cork. Presently he stretches out his neck, and with great exertion of his wings he runs along the top of the water for seventy or eighty yards, until at last, having got sufficient impetus, he tucks up his legs, and is once more fairly launched in the air. It is, probably, this necessity for running along the top of the water before he is able to ascend from it that has given rise to the fable of the albatross being able to walk on the surface of the water with hardly any assistance from his wings, and to the statement that the noise of his tread may be heard at a great distance.

Young bird settling on the water.

The albatross never dives; indeed he never puts his head under water. When caught and placed on deck, he is unable to stand, or to arise unless a strong wind is blowing, and must lie almost helpless on his breast. After albatrosses have been on board for a few minutes they generally, but not always, throw up a large quantity of oil.

The Royal Albatross.

Diomedea regia.

Like the last species, but with the upper back destitute of fine transverse lines; the lesser wing coverts mottled with grey. Eyelids, black. Length of the wing, 25.5 inches; of the tarsus, 4.8 inches. Young, similar to the adult; but the back usually with some irregular brown blotches. Egg—5 inches in length; rarely spotted. Breeds on Campbell Island in the middle of November, the young being hatched in January. It also breeds on the eastern end of Adams Island.

These birds sometimes follow a vessel for days together, and are seldom seen to settle on the water except to feed. Several well authenticated cases are known of albatrosses and Cape pigeons following a vessel for several days in succession when the vessel has been going from 150 to 200 miles in twenty-four hours; but these are exceptional cases. It seems incredible that any animal should be able to undergo so much exertion for so long a time without taking rest; and in these cases it is not perhaps necessary to suppose that they do so. Usually, marked birds are not seen again; but occasionally a marked bird puts in an appearance after having been absent about two or three days. Careful watching has shown that where there were fifty or a hundred birds following a ship in the day, there are only one or two at night, and in the morning-watch they may be seen coming from all sides. The probable explanation, therefore, is that most of the birds sleep on the sea; and in the morning, knowing very well that a vessel is the most likely place in which to obtain food, they fly high, with the intention of looking for one. A few find the vessel they were with the day before, but most find another one. In the latter case, if the second vessel is going in opposite direction to that of the first, they are never seen by the first again. If, however, the course of the two vessels is the same, the bird might lose the second and rejoin the first, after a lapse of two or three days. A height of 1000 feet would enable a bird to see a vessel 200 feet high more than fifty miles off; and often, although unable to see the vessel itself, it would see another bird which had evidently discovered one, and would follow it in the same way as vultures are known to follow one another.

Snowy Albatross.

(*From a Sketch.*)

The Snowy Albatross.

Diomedea chionoptera.

Like *D. regia*, but the lesser wing coverts along the edge of the wing and the proximate middle coverts, nearly pure white; the upper surface of the wing being much whiter than in either of the other two species. Length of the wing, 25.5 inches; of the tarsus, 5 inches. Breeds on Kerguelen Land and Marion Island. It is only stragglers that visit New Zealand. The eggs have a well marked cap of rufous dots at the larger end; their length is 5 to 5.4 inches.

The real home of this species seems to be the neighbourhood of Cape Horn, where it is very common, apparently to the exclusion of the wandering albatross (*D. exulans*).

Genus Thalassarche.

Smaller, with the back dark grey or smoky. Anterior margin of the underside of the wing dark. Southern Ocean.

Key to the Species.

1. Larger: wing 22 inches.	T. salvini.
Smaller: wing 19 to 21 inches.	2
2. Bill yellow.	T. melanophrys.
Sides of the bill dark.	3
3. Lower edge of the bill dark.	T. chlororhynchus.
Lower edge of the bill yellow.	4
4. Yellow of upper edge of bill expanded at base.	T. bulleri.
Yellow of upper edge of bill narrowed at base.	T. culminatus.

Nest of Grey-backed Mollymawk.

The Grey-backed Mollymawk.

Thalassarche salvini.

White, with a pale grey head, except the forehead and crown, which are white. A blackish band in front of the eye and extending over it. Back, wings, and tail, as well as the anterior margin of the under surface of the wing, dark greyish brown. Base of the bill, with a narrow band of black, outside of which is a strip of orange, which does not show on the dried skin. Bill, bluish horn colour; yellow on the upper edge. Legs and feet, bluish white. Length of the wing, 22 inches; of the tarsus, 3.6 inches. Egg—4.25 inches in length.

In some birds the head is nearly white, and in that case they cannot be distinguished from *T. cauta,* of Tasmania. The male is rather larger than the female, and has a stouter bill. They feed on fish, barnacles, and other crustaceans. They breed at the

Bounty Islands, on the rocks.   The nest is a compact cup-shaped structure, about a foot in diameter and five inches high, made entirely of feathers matted together with the droppings of the bird.   In structure the bill much resembles that of *T. culminatus*, but the black membrane behind the nasal tubes is narrower, and is sometimes even absent.

Mollymawk on Wing.   *(Photo. by Mr. G. Buddle)*

The Mollymawk—Toroa.

Thalassarche melanophrys.

White, with black band through the eye.   Back and wings, brownish black.   The tail, dark grey.   Bill, yellowish horn colour.   Legs and feet, yellow.   There is no black membrane behind the nasal tubes.   Length of the wing, 20 inches; of the tarsus, 3.3 inches.   Egg—White, with rufous specks at the larger end; 4 inches in length.

The mollymawk breeds on rocky cliffs on the Campbell Islands, commencing in the middle of September; and it has also been reported from the Auckland Islands. There is no difference between the young and old birds except in the colour of the bill and feet. The former is dark blue, and the legs and feet light blue in the young. The mollymawk sometimes dives, but

(*Photo. by Mr. G. Buddle, taken on Disappointment Island, Auckland Group*)
The Mollymawk.

apparently does not like doing so, generally preferring, when anything good to eat is under water, to let a brown petrel fish it up. It gives chase, and, running along the top of the water croaking and with outstretched wings, it compels the petrel to drop it, and then seizes it before it sinks again. When caught and brought on board, the bird stands pretty firmly on its legs, and does not vomit oil, as most of the petrels do.

The White-capped Mollymawk.

Thalassarche bulleri.

Head and neck, pale grey; a dark patch in front of the eye. Back, wings, and anterior margin of their lower surface, sooty brown. The rest of the body, white. Bill, blackish horn colour, with the upper and lower margins yellow; the yellow of the upper margin expanding backwards. Both in front and behind the nostril there is a narrow band of black membrane, which does not quite reach the base of the bill. Legs and feet yellow. Length of the wing, 20.8 inches; of the tarsus, 3.3 inches. Breeds at the Snares towards the end of January.

(Photo by Mr. G. Buddle, on Auckland Islands.
Mollymawks on Nest.

The Grey-headed Mollymawk.

Thalassarche culminatus.

The colours are the same as in the last species; but the yellow on the upper margin of the bill narrows slightly behind the nasal tubes, and the black membrane behind the tubes is much broader, the two from each side just meeting at the base of the bill; also there is no blackish mark in front of the eye. Length of the wing, 20 inches; of the tarsus, 3.25 inches.

In the young bird the bill is entirely dark, and the head, neck, and lower surface are brown. It is not uncommon on the coasts of New Zealand, but its breeding-place has not yet been ascertained.

(*Gray's Genera of Birds.*)

Grey-headed Mollymawk.

The Yellow-nosed Mollymawk.

Thalassarche chlororhynchus.

Like the last in colour, but the head is white. Bill, black; the upper margin, yellow, passing into orange at the tip; the yellow narrows rapidly behind the nostrils. The black membrane behind the tubes is very broad, and the two from each side unite together broadly at the base of the bill. No black mark in front of the eye. Length of the wing, 18.8 inches; of the tarsus, 3.1 inches.

As in the case of the last species, this bird is not uncommon on the coasts of New Zealand, and the breeding-place has also to be discovered.

Genus Phoebetria.

Sides of the mandible with a deep longitudinal groove. Tail long and pointed. Southern Ocean.

The Sooty Albatross.—Toroa-pango

Phoebetria fuliginosa.

Sooty brown, darkest on the face. Back and abdomen, grey; lightest on the shoulders. Bill, black. The groove on the mandible, blue. Legs and feet, yellow. Young, entirely sooty brown. Length of the wing, 19.5 inches; of the tarsus, 3 inches. Egg—White or buff, plain or with rufous specks, principally at the larger end; the length about 4 inches. The male is larger than the female.

This bird breeds at the Antipodes and the Auckland Islands about the end of October. It makes its nest in inaccessible cliffs, the structure being a conical mound seven or eight inches high, and hollowed into a cup at the top, and the egg is much elongated.

The unrivalled flight of the albatross has been the admiration of voyagers from the earliest times. Those who have watched these birds day after day will agree with Mr. Gould that the sooty albatross should be awarded the highest honours in this respect. Nothing can equal his ease and grace as he sweeps past, often within a few yards, every part of his body perfectly motionless except the head and the eye, which turn slowly and seem to take notice of everything. One of these birds has been watched narrowly, and has been seen sailing and wheeling about in all directions for more than half-an-hour, without making the slightest movement of the wings. Half-an-hour, however, is longer than usual.

This so-called "sailing" method of flight is characteristic of all the larger petrels. It enables the bird to keep on the wing all day with very little exertion. It is not true sailing, but some word is wanted to distinguish it from the soaring of vultures, pelicans, and other birds. For the flight of the petrels is performed near the surface of the sea, and the birds make irregular curves with such sharp turns that their out-stretched wings, when turning, are in an almost perpendicular position. Vultures, when soaring, ascend to a considerable height, and then wheel round and round in great circles, always keeping their wings horizontal.

Sailing flight depends upon the principle of the inclined plane. The bird acquires momentum by flapping its wings, and then holding them extended and motionless, waits until its momentum is nearly exhausted, when it once more propels itself forward as before. In the case of the sooty albatross, the interval may, under favourable conditions, be about half-an-hour, and the

Sooty   Albatross.   *(After Gould.)*

difficulty is to explain why the friction of the air does not sooner bring the bird to a standstill. It was pointed out in 1889, by Mr. A. C. Baines, that the birds usually rise in a slanting direction against the wind, turn round in a rather large circle, and make a rapid descent down the wind. They subsequently take a longer or shorter flight in various directions, almost touching the water. After that comes another ascent in the same manner, followed by another series of movements. As the velocity of the wind near the surface of the sea is diminished by the friction of the waves, when the bird ascends into the more rapidly-moving upper current its *vis inertiae* makes the wind blow past it, and so its stock of energy is increased. When it descends it will be moving faster than the lower stratum of wind, and will again develop new energy if its *inertia* is sufficient to prevent its attaining the new velocity of the wind at once. So that the bird must fly against the wind when ascending, and with it when descending. Thus the energy constantly lost by the friction of the air is partially renewed by these manœuvres. This explains why the birds can sail longer in a high wind than in a calm. It is because in a high wind, and with a high sea, there is much greater difference between the velocities of the wind near the surface and a short distance above it; and this, again, is an explanation of why an albatross keeps so close to the surface of the sea, only just topping the waves and occasionally rising high in the air.

The sooty albatross makes no sound when at sea; but, when breeding, the sitting bird on the nest gives a wailing cry, like ''pee-u, pee-u,'' which is similarly answered by its mate, flying around.

ORDER IMPENNES.

Wings short, covered with scale-like feathers. Tail composed of narrow rigid feathers. Tarsi very short, the anterior toes united by a web; the hind toe very small and united to the tarsus Southern Hemisphere.

It is stated in the Introduction that New Zealand, together with the neighbouring islands, may be looked upon as the headquarters of the penguins, as all the genera except one, *Spheniscus*, are found here. Besides this, there is evidence that points to New Zealand being the centre of dispersion. The oldest penguin known, *Palaeeudyptes antarcticus,* is from the Eocene or Oligocene rocks of New Zealand. It is a true penguin, and except that the wing is proportionately longer than in living penguins, it shows no intermediary character. The only other known fossil penguins are four species of *Palaeospheniscus,* and of *Paraptenodytes,* from the Miocene rocks of Patagonia. It is also worthy of note that remains of *Palaeeudyptes* are found in the Oamaru freestone, which represents an old coral reef. In these circumstances, it is fitting that the penguins should be given some latitude in regard to space in this New Zealand work.

Evidently, they are specially adapted for an aquatic life. On account of this adaptation, and not by degeneration, they have lost the power of flight. The feathers on the wings and on the body have been reduced until they are almost like scales, which form a continuous covering over the body. The only fairly long feathers with quills are in the tail, or form plumes on the head; in the wings there are no quill feathers.

They can dive better than any other birds. Under water, their wings are used as a means of locomotion, their feet being stretched straight out behind. They fly through the water as other birds fly through the air. As, however, the bird is lighter than the water, the principal stroke of the wing must be directed upwards, so as to keep the body under the surface. For this purpose, the muscles that give the up-stroke to the wing are developed to a greater extent than in other birds, and there is a consequent expansion of the scapular, a bone that springs from the shoulder of the wing, and lies along the side of the back. As the wing is reduced to a flipper, the bones have all become modified, and are broader and flatter than in other birds, and the first digit has become fused with the second.

These modifications have turned the wings into strong swimming paddles. So rapid is the flight of these birds under water that the crested penguins and the rock hoppers spring out of the water, with their wings close to their sides, and take long leaps through the air like porpoises.

Penguins are the only birds that can swallow their food under water. To do this, however, they have not developed any structural modification. It is constant practice, not special adaptation, that enables them to remain under water longer than other birds can. The limit of their power to remain under, without coming up to breathe, is not known. Sir John Murray, of the Challenger Expedition, states that a crested penguin placed in a basket and submerged was dead in a minute and a half. This, however, can hardly be taken as a fair test. A peculiarity in the skeleton of a penguin is the shortness of the three metatarsal bones of the leg, which are separated by deep grooves throughout their whole length, the grooves generally penetrating completely through the bone in two places, so as to produce two large intermetatarsal foramina.

Although these birds are naturally not so active on land as in water, it is a mistake to suppose that they are plantigrade, in other words, that they apply the lower surface of the metatarsus to the ground when walking or hopping. They walk or hop on their toes like other birds, and it is only when resting that they place the metatarsus on the ground, a habit which is by no means peculiar to penguins. It is stated that on the snow or smooth ice they lie down on their stomachs and push themselves along with their wings so rapidly that a man running can hardly keep up with them.

On shore, they sleep a good deal in the daytime, tucking their heads behind their small wings, and at night they make a hideous noise. Night and day are much the same to them. They feed largely on cuttle-fish and crustacea; but, no doubt, they eat fish as well. In size and colour, both sexes are alike, but the young birds generally differ from the adults.

The usual number of eggs in a penguin's nest is two, but the rock-hopper lays its two eggs at about two months' interval, the second being laid after the first is hatched, so that it has really two broods of one each. Members of this species make no nest, but carry their egg between their legs. The king penguin has improved upon this. It also has only one egg at a time, but it carries this egg in a fold of skin between the legs, so that the egg is quite hidden. The egg also differs in shape from the egg of other penguins in being pointed at one end, which must help its being retained in the fold of skin.

The species of *Catarrhactes* makes rough nests of grass, where any is to be obtained. Each parent sits on the eggs alternately. When the time for changing comes, the relief places itself close alongside its mate and pushes it off the nest, covering the egg at once. This is done to prevent the sea hawks, which are always hovering round, from eating the eggs. The yellow-eyed penguin breeds under logs of wood or among tussock-grass, only a few pairs living together; while *Eudyptula* and *Spheniscus* make their nests in holes.

During the breeding season, the king penguin, as well as the species of *Pygoscelis* and *Catarrhactes,* collect in large numbers, and form rookeries, which have often been described; but the other species do not do so. There are no large rookeries on the Auckland or Campbell Islands, nor among the West Coast Sounds of New Zealand. *Spheniscus* breeds in holes, as a defence, no doubt, against predaceous mammals, in South Africa, South America, and Australia; but that cannot be the object in New Zealand. It is only those genera living on islands, or on the Antarctic Continent, that form rookeries; and there is an intermediate stage, represented by *Megadyptes,* and sometimes by *Catarrhactes,* which inhabits islands, but does not collect together in large numbers. From this, it may be inferred that the habit of forming rookeries is a late one, induced probably as a defence against seals.

Penguins that live in rookeries spend about eight months on or near the shore, and four months at sea, when the breeding grounds are quite deserted; but they do not seem to go far away, as they are rarely seen more than fifty miles from land.

The times of breeding are not yet well known. The blue penguin is the first, and commences in the early part of September, while the white-flippered penguin does not begin until October. In the Antipodes and Bounty Islands the big-crested penguin comes to the land in September, the tufted penguin arriving at the Antipodes a little later; and they leave in May or June, when the young are sufficiently strong to undertake a sea voyage. Dr. H. Filhol says that in Campbell Island the tufted penguin begins to lay early in November, while in Kerguelen Island it does not arrive until then, and commences to lay about December. In the islands of St. Paul and Amsterdam they are said to be much earlier, arriving in July, and leaving for the sea in March. The royal penguin, in Macquarie Island, commences to lay early in November. The rock hopper is an early bird, and at Kerguelen Island the young of the first brood hatch out towards the end of October, and the second in December. The yellow-eyed penguin breeds in New Zealand at the end of October; but in Campbell Island, according to Dr. Filhol, its eggs are laid early in November, the young being hatched out by the end of the month. The king penguin, at Macquarie Island, commences to lay its eggs about the middle of November.

The penguins differ greatly in their dispositions. *Megadyptes* and *Catarrhactes* are sulky in captivity, while the king and the little blue penguin are much more friendly, the former especially being easily tamed.

Penguins belong entirely to the Southern Hemisphere. To the south they extend as far as the Antarctic continent; while on the west coast of South America they are found as far north as the coast of Peru, and one species inhabits the Galapagos Islands, which are situated on the Equator. On the east coast of South America they extend only to Rio Grande do Sul. In Australia and South Africa they inhabit the southern coasts only. They are a small group of birds forming a single family. But this family may be divided into three sub-families, which have slightly different geographical distributions. The first sub-family contains the genera *Aptenodytes* and *Pygoscelis,* which inhabit the Antarctic Regions from the ice up to about 53° S. The second sub-family consists of the genera *Catarrhactes* and *Megadyptes.* These are found between 55° S. and 38° S. The third sub-family contains *Spheniscus* and *Eudyptula,* which do not live further south than 45° or 50°, and extend to the most northern limits reached by the penguins.

They have descended from flying birds. This is proved by the structure of the wing. The bones are on the same pattern as that found in other birds. In addition to this, several of the muscles of flying birds are represented in the penguins by non-contractile tendinous bands, which are functionally useless, but have not yet altogether disappeared. It is certain that they are not closely related to the auks of the Northern Hemisphere, which are somewhat like the penguins in appearance, but that they come nearest to the petrels, or tubinares, although the two groups are so different in form. This makes it difficult to guess what the ancestors of the penguins were like.

Of the genera of living penguins, *Pygoscelis* connects *Aptenodytes* with *Megadyptes,* and this last genus may be taken to represent the ancestral form of the other two. *Spheniscus* is a specialised form of *Eudyptula.* So, if we try to draw up a genealogical tree of the penguins, we must place *Megadyptes* and *Eudyptula* near the base, and make *Aptenodytes, Catarrhactes,* and *Spheniscus* occupy the apices of the branches. Paleontology points to *Megadyptes* as the oldest living form; but Mr. Pycraft is of opinion that *Eudyptula* lies nearest to the ancestral stock.

As to the colour of the earlier forms of penguins, it may be supposed that they were white below and dark on the back, head, and throat. This last is inferred to have been the case, because in *Eudyptula* the young in the down has the head and neck brown, and it remains brown throughout life in all species of *Spheniscus.* It is probably the same in *Megadyptes,* for in young birds the feathers on the throat are brown, like those of the head. In the royal penguin, also, the chin and throat are brownish in the young. In all these cases, except *Spheniscus,* the chin and throat subsequently become white, which, therefore, must be a late acquisition. In the other species of *Catarrhactes* the young birds in the down have the throat black; then, with the first feathers, it gets grey or nearly white, and then black again. Also the species of *Pygoscelis* have the chin and throat white when young, but black when old. From all this we may infer that the black throat in these species is due to atavism, and that their immediate ancestors had white throats when adult. We may, therefore, suppose that the royal penguin represents the earliest species of the genus *Catarrhactes,* and from it the others have descended; and we may further suppose that the tufted penguin has been developed from the crested penguin, because the plumes on the head are so much longer; so that we have the yellow-eyed penguin, the royal penguin, the crested penguin, and the tufted penguin as the probable line of descent.

Another interesting point with regard to the penguins is their migrations, and the directions these have taken. New Zealand is the only district in which both *Megadyptes* and *Eudyptula* live. Formerly, it was the home of *Palaeeudyptes*. Therefore, it is reasonable to assume that New Zealand was the centre from which the birds dispersed. Some explanation of this is afforded by geological evidence. During the Eocene Period the Andes had not been formed, and the highlands of Tierra del Fuego were not connected with the lands of the Northern Hemisphere; and South Africa seems to have been isolated from Northern Africa. Therefore, Australasia was the only district where the land of the Northern Hemisphere pushed down far into the Southern Hemisphere.

The next point that arises is in respect to the direction of the migration from New Zealand. The species of *Pygoscelis*, *Aptenodytes*, and the tufted penguin, being spread round the Southern Ocean, do not help in a solution of this question. Had the ancestors of *Spheniscus* passed from New Zealand westward to South Africa, it would be reasonable to expect to find their descendants on some of their immediate islands, such as St. Paul, or the Crozets. This, however, is not the case. Therefore, it seems to be more probable that they travelled eastward to South America, where they arrived in Miocene times, and thence to South Africa.

In the Southern Ocean the current runs from west to east, and this would favour the spread of the birds towards the east. It is evident that a penguin could swim across the Pacific Ocean provided it could obtain food on the way, and Sir James Ross saw some nearly 1200 miles from the nearest land. It has been supposed that in their migrations they have been much helped by icebergs, as penguins have been seen sitting on them. But icebergs generally drift to the north, and could not supply the birds with food. On the contrary, the melting ice would probably destroy or drive away any fish or crustaceans in the neighbourhood. At first it is surprising to learn that *Spheniscus* should have spread so far north as the Galapagos Islands, but this is easily explained when it is noted that the cold Antarctic current, sweeping up the coasts of Chili and Peru, reduces the temperature of the ocean at the Galapagos to 62° or 66° F., whereas its normal temperature at the Equator is 81° to 88° F.

There is yet one more point of interest in connection with penguins. What part has natural selection played in their development? Their feathers, their wings, and the backward position of the legs are obviously adaptations for an oceanic life, and the fatty subcutaneous layer is an adaptation for keeping the body warm. All these are due to natural selection.

But it cannot be said with certainty that all the generic characters are adaptations. The differences here are chiefly in the bill and the tail. In *Aptenodytes,* the bill is long and slender and is curved downwards at the tip. This seems to be an adaptation for catching fish; but it has not been stated that the king penguin feeds more on fishes than other species do. The chief character of *Catarrhactes* is the strong bill, which is swollen at the base of the latericorn; and, as the males have larger bills than the females, it may be supposed that this is an adaptation for fighting, and it may be due to natural selection. But it is hard to suggest a use for the feathers on the bill of the *Pygoscelis,* or for the longitudinal grooves at the base of the mandibles in *Spheniscus.* In *Eudyptula, Spheniscus, Megadyptes,* and *Aptenodytes,* the tail is short, and is composed of from sixteen to twenty feathers; in *Pygoscelis* and *Catarrhactes,* it is long, and is composed of from twelve to sixteen feathers. Of what use can these differences be? Penguins may, perhaps, use their tails as rudders, but it is difficult to say which of these different tails would answer the purpose best. It must be remembered that different genera of penguins sometimes inhabit the same island, as at Kerguelen, the Falklands, and Macquarie Islands. They seem to have the same food, and the same methods of capturing it. Genera which inhabit similar localities when breeding, and which feed together, so far as is known, on the same kind of food, could not have been differentiated by the direct action of external conditions, and yet it is equally hard to explain how this could have been brought about by natural selection.

T

The specific characters are chiefly the differences in the colour of the plumage or of the eyes, and some of these characters are probably due to that part of sexual selection which has been called preferential selection. The long plumes of the tufted penguin and the yellow bands on the king and emperor penguins may be thus accounted for, while the differences in the species of *Spheniscus* may possibly be recognition marks. But there are some exceptions. The black throat of four different species of *Catarrhactes* and two of *Pygoscelis* cannot be due to sexual selection, because we cannot suppose that six different species belonging to two different genera all wished to change their white throats to black ones at the same time. This seems to be due simply to reversion; but it is a very interesting case. It is possible that the red eyes of the tufted penguin may be a recognition mark to distinguish it from the brown-eyed crested penguin. It is possible that the earliest members of the tufted penguins were driven from the rookery on account of their red eyes, and in this way they have been forced to keep together; but it does not seem likely that the white marks on the wings of the big crested penguin and the white-flippered penguin are recognition-marks, for they breed in different localities from the crested penguin and the blue penguin, from which they were derived.

The effects of isolation are not so well marked in the penguins as in most birds, owing, probably, to their wandering habits, and to the difficulty they must experience in returning after a long voyage to the place from which they started. Still we find that the royal, the crested, the big crested, and the white-flippered penguins, and the four species of *Spheniscus,* all inhabit separate localities. It might be expected that the tufted pengin, being so widely spread, would show more decided variation than it does. Differences, however, do exist. Mr. Watson has shown that the skull of the birds inhabiting the Falkland Islands is larger than those of birds found on Tristan d'Acunha, and that these again are larger than those of birds from Kerguelen Island, while the Falkland Island birds have a smaller bill than that of any of the others. The birds inhabiting St. Paul and Amsterdam are bluer in colour, and have longer head plumes, than those from New Zealand or the Falkland Islands.

Altogether, it is reasonable to conclude that natural selection has been instrumental in forming the family characters, and that sexual selection has been the cause of some of the generic and specific characters, but that there are other generic and specific characters which are not due either to one or the other. At the same time, there is not the least reason for supposing that any of the characters are due to the action of environment.

It may be added, as an item of general interest, that the name penguin was originally given by Spanish sailors to the short-winged northern auks and divers, owing to the quantity of fat (*pinguie*) found on them. On the discovery of the Southern Ocean, the name was used for the somewhat similar birds found there. Subsequently, the name was dropped for the northern birds, and was retained for the southern birds only. The French still call the auks *pingouin* and the penguins they call *manchot*.

Key to the Genera.

1.	Lower mandible curved downward.	Aptenodytes.
	Lower mandible straight.	2
2.	A yellow band over the eye.	3
	No yellow band over the eye.	4
3.	Yellow feathers long, but not passing to back of head.	Catarrhactes.
	Yellow feathers short, passing to back of head.	Megadyptes.
4.	Feathers advancing along sides of mandible.	Pygoscelis.
	Feathers not advancing along sides of mandible.	Eudyptula.

Genus Aptenodytes.

Large birds with the bill long and curved downwards towards the tip. Tail rather short, composed of twenty feathers, which are almost hidden by the upper tail coverts. Tarsi covered with feathers. Antarctic Seas.

The King Penguin.

Aptenodytes patagonica.

Above, bluish grey; below, white. A patch of yellow on the sides of the head, continued in a band down each side of the throat, and uniting with one another on the fore part of the neck. Bill, black; the base of the lower mandible pink. Legs and feet, black. Eye, bright brown. Total length about 36 in. Egg—4 in. in length; pointed at one end. Islands of the Southern Ocean.

The king penguins form large rookeries at Macquarie Island, but do not breed on any other of the New Zealand group. They make no nest, and lay only one egg, which the female holds in a fold of the skin between the legs until it is hatched.

Rock   Hopper.   (*Voy. Erebus and Terror*)

Genus Pygoscelis.

Bill moderately long or short, not very stout; the feathers of the sides of the face advancing along the lower mandible. Tail long, composed of 12 to 16 feathers; the upper tail coverts short. Antarctic Seas.

The Rock Hopper.

Pygoscelis papua.

Above, slaty grey. The head and neck, brownish black. A white band across the crown. Below, white. Eye, rich brown. Total length about 30 in. Length of the egg about 2.75 in. In the young the chin and throat are white. Breeds on Macquarie Island.

Dr. Kidder, in dealing with this species, graphically describes its movements. "No living thing that ever I saw," he writes, "expresses so well a state of hurry as a penguin when trying to escape. Its neck is stretched out, flippers whirring like the sails of a windmill, and body wagging from side to side, as its short legs make stumbling and frantic efforts to get over the ground. There is such an expression of anxiety written all over the bird; it picks itself up from every fall, and stumbles again with such an air of having an armful of bundles, that it escapes capture quite as often by the laughter of the pursuer as by its own really considerable speed."

Genus *Catarrhactes*.

Bill moderately long and very stout, the sides of the upper mandible much swollen near the base. Tail long, composed of 14 or 16 feathers; the upper tail coverts short. Southern Seas.

Key to the Species.

1. Yellow bands uniting on forehead.	C. schlegeli.
Yellow bands not uniting.	2
2. Smaller, yellow feathers very long.	C. chrysocome.
Larger, yellow feathers shorter.	3
3. White edge on flipper narrow.	C. pachyrhynchus.
White edge on flipper broader.	C. sclateri.

The Tufted Penguin.

Catarrhactes chrysocome.

Above, slaty grey, the head and throat darker; a band of golden feathers over each eye, the posterior of which are much elongated and drooping. Below, white. Flipper edged posteriorly with white. Eye, red. Total length about 25 inches. Young birds have the chin and throat greyish white. Egg—2.5 inches in length, pale blue. Islands of the Southern Ocean. Breeds on the Snares, Antipodes, Auckland, Campbell, and Macquarie Islands, but not in great numbers.

The bird is sometimes called the Victoria penguin by New Zealanders.

(*Challenger Report.*)
Tufted Penguins: old and young.

The Crested Penguin.—TAWHAKI.

Catarrhactes pachyrhynchus.

Like the last in colours, but the yellow eye-stripe not so much elongated, and the feathers not drooping. Posterior margin of the upper surface of the flipper with a very narrow white margin of one row of feathers only. Eye, brown. Total length about 28 in. In the young the chin and throat are greyish white. Egg—2.85 in. in length. New Zealand only. Forms large rookeries on the Snares, and breeds in small quantities on the southern coasts of New Zealand. (For a figure of this bird see *Nature in New Zealand*, p. 38.)

The crested penguins breed in small colonies up to twenty-four together. After hatching, the female remains with the young for the first few days, and the male brings food, which consists of fish, for all. When they are not disturbed, they walk or hop upright rather clumsily; but when they are startled, they stoop down and use their flippers as fore-legs. For climbing up the rocks they use their bills, and then they get along very quickly.

(*Buller.*)

Big Crested Penguin.   Blue Penguin.

In swimming, the body is under water and only the head out. They swim slowly; but when they dive, they go with great rapidity.

"This noble bird," to quote Mr. Reischek, "has been found on the coast of the South Island, but is most plentiful in the West Coast Sounds, especially Dusky and Milford. In Dusky Sound there are several colonies, two in Supper Cove and one on Cooper's Island. These birds come on shore in July, when they begin to build their nests, which consist of a few sticks and leaves, which the male brings while the female constructs a careless nest, either in a cave between cliffs or under large stones, in which she lays one or sometimes two eggs.''

(*Photo by Mr. G. Buddle, on Antipodes Island.*)
Big Crested Penguin on egg.

The Big Crested Penguin.

Catarrhactes sclateri.

Like the last, but the white margin on the posterior margin of the flippers broader, and composed of two rows of feathers. Eye, brown. Total length about 29 in. New Zealand only. Forms rookeries on Antipodes and Bounty Islands. Beak, brownish red; feet, flesh-colour. Breeds in September.

The Royal Penguin.

Catarrhactes schlegeli.

Above, slaty grey; below, including the chin and throat, white. The yellow eye-stripes meet in front, forming a broad yellow frontal band. The bill also is much longer than in the other species. Total length about 30 in. In the young the throat is grey. Egg—3.15 in. in length. New Zealand only. Forms large rookeries on Macquarie Island.

Genus Megadyptes.

Bill moderate in length and in stoutness, the upper mandible only slightly swollen at the base. Tail short, composed of twenty feathers; the upper tail-coverts short. New Zealand islands only.

The Yellow-eyed Penguin.—HOIHO.

Megadyptes antipodum.

Above, slaty grey: below, including the chin and throat, white. Crown encircled by a golden band of short feathers. Eyes, lemon yellow. Total length, about 30 in. In the young, the yellow band is confined to the sides of the head. Breeds early in September in small quantities on the southern coast of New Zealand, Stewart Island, Auckland and Campbell Islands. It is sometimes called the Grand Penguin.

Genus Eudyptula.

Small birds, with a stout bill. Tail very short, composed of sixteen feathers, which are entirely concealed by the long upper tail-coverts. New Zealand, Tasmania, and South Australia.

The Blue Penguin.—KORORA.

Eudyptula minor.

Above, slaty blue; below, white. Flippers greyish black, rather narrowly edged with white on the posterior margin. Eye silvery grey. Total length about 16 in. The young resemble the adult. Egg—2.25 in. in length. Found in Tasmania and South Australia as well as New Zealand. Here it occurs from the North Cape to Stewart Island, and at the Chatham Islands.

The blue penguin breeds in caves, or in burrows, sometimes twelve feet long. The burrow is about three or four inches in diameter, and in it the females, about the end of September, lay two eggs. Young birds appear in November; and, at the end of February, or in March, they leave the shore, returning early in September. The female is smaller than the male. They feed upon fish and crustaceans. They walk badly, shuffling along with the body thrown forwards, with a curious undulating motion.

The White-flippered Penguin.

Eudyptula albosignata.

Like the last, but both margins of the flipper widely bordered with white, and a more or less distinct white patch near the middle of the posterior margin. Eye, silvery grey. Total length, about 16.5 in. Egg—2.15 in. in length. At present known only from Banks Peninsula.

These birds may be found breeding in the months of November. December, and January. "They nest in large numbers, amongst crevices of rock usually not far above high-water mark," says Mr. Potts, "so as to have immediate access to the sea; and these may be considered as choice stations for rearing their young. Perhaps less fortunate, or more industrious, couples have to undertake a far greater amount of toil in the accommodation of a family. They burrow out a tunnel with very great neatness, often for a considerable distance from the entrance, which is usually a perfectly round hole some three or four inches in diameter. When the tunnel is molested, the old bird makes a vigorous defence of its offspring, using beak and claws with much spirit, at the same time uttering cries not very unlike the mewing of a cat."

ORDER STEGANOPODES.

All four toes connected together by a broad membrane. Throat furnished with a pouch.

Key to the Genera.

Bill pointed at the tip.	Sula.
Bill hooked at the tip.	Phalacrocorax.

Genus Sula.

Bill stout, pointed; the nostrils completely closed in adults. Wings very large and pointed, the first quill longest. Tail wedge-shaped. Claw of the middle toe serrated. Temperate and tropical seas.

The Gannet.—TAKUPU.

Sula serrator.

White, with the top and sides of the head buff. Quills and four middle tail feathers black. Bill horn-colour, tinged with blue. Legs and toes greenish yellow, the webs brown. Eye pale silvery brown. Length of the wing, 18 in.; of the tarsus, 2.2 in. The young have the upper surface brown spotted with white. Egg—White, length 3 in. Australia and New Zealand.

Gannet.   (*Meyer.*)

The gannet is common in the northern parts of New Zealand, but is rarely seen south of Cook Strait. It breeds on the Great Barrier Island; White Island, in the Bay of Plenty; and Gannet Island, off Kawhia. The nests are placed very close together, and are roughly composed of grass and seaweed. It is a very awkward bird on land, but an excellent flier, soaring over the sea, and darting down, with closed wings, on any fish that may be near the surface, disappearing under the waves, and coming up again at some little distance with its prey in its mouth. It is remarkable that it should not occur at the Chatham Islands.

Cormorants.

Genus Phalacrocorax.

Bill straight, rather slender, hooked at the tip; the nostrils rudimentary. Wings moderate, pointed, the second and third quills longest. Tail rounded or wedge-shaped. Claw of the middle toe pectinate. The whole world, except Polynesia.

New Zealand waters contain more different kinds of cormorants, commonly called shags, than the waters of any other part of the world. While we have fifteen species, there are only twelve in North and South America, seven in Asia, six in Africa, five in Australia, and three in Europe. The large number of species in New Zealand is attributed to two causes. First, this country was the meeting-place of two migratory streams, one from the Malay Archipelago and New Caledonia, the other from Patagonia. Secondly, New Zealand has been broken up into a number of islands, lying at considerable distances apart, and these have been isolated for a very long time.

Cormorants are found all over the world, except in Polynesia, east of New Guinea, the Louisiade Archipelago, and New Caledonia. This is surprising, as fish, on which these birds feed, are very abundant round the Polynesian Islands. Another remarkable fact is that all cormorants from all parts of the world are considered to belong to one genus, namely, *Phalacrocorax,* which, however, can be divided into several sub-genera. Most cormorants live amongst rocks on the sea-shore, but some prefer rivers, and live far inland, and these generally build their nests in trees, considerable numbers consorting together.

They are greedy birds, and display remarkable dexterity and boldness in pursuing and seizing fish. Sir Julius von Haast, writing of the pied shag, gives a good illustration of the diving and fishing capabilities of cormorants generally. He was standing near a spot where one of the northern spurs of Mount Murchison slopes down to the Buller River, which there forms small falls and rapids. A cormorant was standing on an isolated rock, round which the foaming waters dashed. By-and-by the bird suddenly jumped into the white foam, to the great surprise of the observer. "In the first instance," he says, "I thought he would not get out again, but would be dashed to death by the whirling waters; but soon he re-appeared, swimming rapidly towards the edge, and then flying on to his old observatory to continue his sport. It is probable that small fishes are taken down by the falls, and, being stunned by the force of the water, are easily caught by the courageous bird."

New Zealand's cormorants afford a good field for a test of current theories in regard to variation, and it will not be out of place to treat the subject briefly in these pages. The intention is not to make a close and formal investigation of the species, as the extent of the present knowledge of the habits and changes of plumage of the birds is not sufficient for the purpose, but merely to give a slight sketch, showing the way to a more elaborate study at some future time.

Although everyone admits that the different kinds of animals have had common ancestors, it is acknowledged that to draw up a genealogical tree with any approach to accuracy would be a very difficult task. No one, perhaps, can fully realise the greatness of the difficulty unless he has tried to reconstruct some portion of the tree. The temptation to undertake the task, however, is very strong, because, if we could trace out the history of an order, of a family, or even of a genus, we should at once obtain interesting information about the origin of variations. The best plan is to find a group of animals in which all the conditions of life are as simple as possible. By doing this, many difficulties and uncertainties will be removed. Conditions are simpler among the sea-birds than among the land-birds, and New Zealand is particularly well adapted for the study of the former. Therefore, our cormorants are specially well suited for this investigation.

They may be divided into three groups. The first consists of five species, three of which are also found in Australia, and they can be distinguished by their black legs and feet. The second group contains only two species, both endemic, which have yellow or orange legs and feet, and a double crest on the head. The third group consists of eight species, all peculiar to New Zealand seas; and these have pink or reddish legs and feet, and the crescent is either single or absent.

The first group evidently comes to us from the north. The pied shag, which is one of the two endemic species, is closely related to *Phalacrocorax gouldi* and *P. hypoleucus* of Australia. The second endemic species is *P. brevirostris*. This and *P. melanoleucus*, which occurs in the Moluccas, Australia, and New Caledonia, are so closely related, and have so many intermediate links, that some naturalists are inclined to consider them as varieties of one species. The geographical distribution of the birds leads to the belief that *P. brevirostris* is descended from *P. melanoleucus*, and the latter from *P. pygmaeus* of the Mediterranean and Central Asia. There is confirmatory evidence of this in the fact that *brevirostris* is a very variable species, apparently not yet capable of breeding truly, while *P. melanoleucus* is not variable, but is an old and well established species.

The curious point in regard to this question is in connection with the plumage of *P. brevirostris* and that of its ancestors. *P. pygmaeus* is black, with scattered white plumes on the head, back, and abdomen; *P. melanoleucus* is black above and white below, with a white frill on each side of the neck; but *P. brevirostris* is black, with a white chin and throat. If, therefore, the evidence from geographical distribution, and from variability, can be trusted, the descendants of *P. pygmaeus*, as they travelled southwards, acquired a pure white breast and abdomen. But when some individuals reached New Zealand, there was a new change, the breast and abdomen becoming black.

Some ornithologists would explain these variations in *P. brevirostris* as cases of partial albinoism, or of intercrossing with *P. melanoleucus*. But if the variations are due to albinoism, it might be expected that the white would appear on the upper as well as on the lower surface, and that it would be unsymmetrical. As to the other theory, *P. melanoleucus* is not sufficiently common in New Zealand to allow us to suppose that the abnormal birds are hybrids. A simpler explanation of these variations is found in the theory of occasional reversion to ancestral characters. Evidently, the surrounding conditions are not sufficient to account for the origin of the variations in *P. brevirostris*. If they were, other cormorants, with white breasts and similar habits, like *P varius,* would be affected in the same manner. It seems more probable, therefore, that the variation, as suggested before, is due to a partial reversion to *P. pygmaeus,* and that the reversion is not yet complete, so that the plumage of *P. melanoleucus,* the immediate ancestor, often appears.

Supposing that the variation has arisen from reversion, it may be asked, how has it been preserved? Cormorants had no enemies in New Zealand, and a black abdomen cannot be better suited than a white one for fishing, as many cormorants are white on the lower surface. Natural selection, therefore, is out of the question. The idea of sexual selection is not satisfactory, and, before forming an opinion on this point, we must know a great deal more about the breeding habits of these birds. It should also be stated, before passing from *P. brevirostris,* that, when young, it is entirely black, the white chin and throat coming afterwards. Why should there be a partial advance in the white throat after a commencement has been made with the entirely black plumage of *P. pygmaeus?* No explanation of this can be offered at present, though it should be pointed out that a white throat is a feature of the young of other species.

The second group of New Zealand cormorants, *P punctatus* and *P. featherstoni,* forms the sub-genus *Sticticarbo,* in which the South American species, *P. gaimardi,* is also included. The facts are too scanty to justify an opinion as to which way this group has travelled. It may be possible to learn something about the migrations of the third group. It forms the sub-genus *Leucocarbo,* and, besides the New Zealand species, contains four from South America and one from Kerguelen Island. Our species of the sub-genus may be divided into a carunculated section,

containing *P. carunculatus, P. onslowi,* and *P. traversi,* and a
non-carunculated section, containing *P. stewarti, P. campbelli,
P. colensoi, P. ranfurlyi,* and *P. chalconotus.* As the skin on the
lores is at first smooth, and then becomes granulated, it may
be supposed that the carunculated section is descended from the
non-carunculated one. Many of the species are characterised by
having white bars on the wing-coverts and lower back, which
do not appear until the birds are mature. Consequently, it may
be assumed that the species with these white bars are descended
from those without them. All the carunculated species, except
P. verrucosus, from Kerguelen Island, have white bars on the
wings, and all but *P. verrucosus* and *P. traversi* have white dorsal
bars. This is confirmatory evidence of the carunculated section
having descended from the non-carunculated one.

Of the non-carunculated species, *P. stewarti* is the only one in
which both bars are present, and it must, therefore, be looked
upon as the connecting link between the sections. In *P. colensoi*
and *P. campbelli* there is an alar bar but no dorsal one and in
the latter bird the alar bar is very narrow. In the South
American species, *P. bongainvilli,* and *P. magellanicus,* there is
no alar bar. So that we have a series from *P. stewarti* to *P.
magellanicus.* The same thing is noticeable in the colouration of
the throat and neck. In *P. magellanicus* both are dark; in *P.
bongainvilli* and *P. campbelli* the throat is white and the neck is
dark; in *P. colensoi* both neck and throat are white; finally, the
young of *P. colensoi* and *P. ranfurlyi* have the neck black as in
P. campbelli, and it becomes white only when the bird is mature.
There is pretty good evidence, therefore, that *P. magellanicus*
represents the prototype from which the others have come.

P. chalconotus is entirely black, but, as the colours of the skin
on the face and gular pouch are exactly like those of *P.
stewarti,* it may be looked upon as a black descendant of that
species, especially as it has occasionally white feathers on its
lower surface.

The carunculated species from South America, *P. atriceps* and
P. albiventer, as well as the Kerguelen Island *P. verrucosus,* differ
from New Zealand forms in having no feathers on the gular

pouch, and they must be considered as descendants of New Zealand forms, as they are more specialised. If this is correct, the white bars on the wings and back must have been lost by *P. verrucosus*.

If there is any truth in these speculations, it follows that our third group of cormorants came into the Southern Ocean by South America, whence they spread to New Zealand. Here they underwent considerable alteration, and the altered forms re-migrated to South America, and some, at last, found their way to Kerguelen Island. When these wide migrations took place, there was probably more Antarctic land than there is at present, but it must have been in the shape of islands, otherwise the South American land birds would have migrated with the cormorants.

The following is the supposed genealogical tree of the sub-genus *Leucocarbo*:—

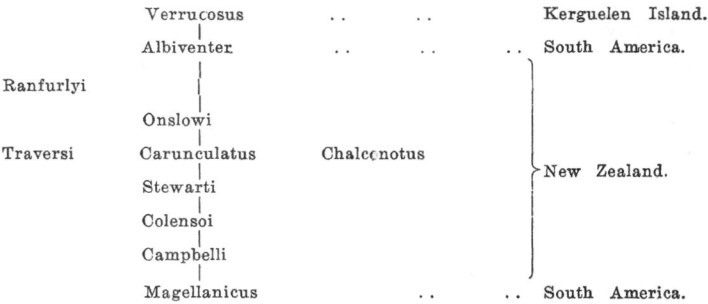

	Verrucosus	.. ..	Kerguelen Island.
	Albiventer	.. ..	.. South America.
Ranfurlyi			
	Onslowi		
Traversi	Carunculatus	Chalconotus	
	Stewarti		New Zealand.
	Colensoi		
	Campbelli		
	Magellanicus	..	.. South America.

P. carunculatus has lost its crest, *P. chalconotus* and *P. traversi* have lost the white dorsal bar, and *P. ranfurlyi* has lost caruncles, crest, and dorsal bar.

Coming back to the origin of the variations, it may be asked, "How did they arrive?" The white bar was a new character acquired by *P. colensoi*, and the white dorsal bar another new character first acquired by *P. stewarti*. It is impossible to suppose that they were caused by amphimixis, as the blending of the sperm and the ovum could not have produced characters that never existed in any of the ancestors of either parent. Nor can the white bars be attributed to the action of the environment, for it is impossible to connect the origin of white feathers on the wings and back with the weather, or with any of the surrounding

U

objects, especially as in *P. chalconotus* the changes have been in the opposite direction; and not only have the white alar and dorsal bars been lost, but the whole of the under surface has also turned black. *P. chalconotus* and *P. stewarti* live together in Stewart Island, and even inhabit the same shaggeries. It is, therefore, impossible for these opposite variations to have been caused by external conditions.

The origin of the white alar bar in *P. campbelli,* or the white dorsal patch in *P. stewarti,* being new characters, cannot be due to reversion. Can they be accounted for by sexual selection? What right have we to assume that a preference was shown by individuals of the opposite sex to one that had some white feathers on its wing or back? Even if this was the case, what guarantee is there that successive generations would all show the same preference? Unless they did do so, the selection would be destroyed, and the variation would not accumulate. The bright colour of the skin on each side of the face in cormorants will be looked upon as a typical example of sexual selection; but the same difficulty occurs here also. Why should both sexes prefer the same colours in their partners as they have themselves, although they cannot see their own colours? Why should the preference for one particular colour, as for crimson in *P. onslowi,* go on for generation after generation in one species, while another species has a similarly constant predilection for another colour? Why did *P. ranfurlyi,* on Bounty Islands, and *P. traversi,* on Macquarie Island, lose their white dorsal patches? Why did *P. carunculatus,* in New Zealand, lose its crest? Can the loss of a white patch, or a crest, as well as the acquisition of those characters, be put down to sexual selection?

The loss of the white bar and of the crest is also probably due to reversion, and other specific characters may have the same origin. New characters, however, which were not due to reversion, occasionally arose. When this took place on continental areas, we may, perhaps, attribute their preservation to their usefulness as recognition marks. But this will not help us with those species found only on a single island, or in a single locality far away from other species. The difference in the external conditions of life for the cormorants cannot be so great in New Zealand as through America from Canada to Tierra del Fuego, or as in Africa and Asia taken together; and yet there are fifteen species in New Zealand to twelve in America, and thirty in Asia and Africa combined. Consequently, we cannot suppose that their specific characters depend entirely upon external conditions.

All the generic characters of the group, such as the hooked bill, the rudimentary nostrils, the close, glossy plumage, the short legs, and the large webbed feet, are eminently adapted to the birds' mode of living. In comparing the sub-genera, however, we find characters which cannot, at present, be set down as useful. The shape of the bill, for example, varies a good deal, being long and slender in *Sticticarbo* (*P. punctatus,* etc.), and comparatively short and stout in *Microcarbo* (*P. melanoleucus,* etc.), while the birds do not show any difference in habits. In *P. campbelli,* on the contrary, there is a difference in habits without any corresponding modification. In the sea around Campbell Island there are hardly any fish, and, according to Dr. Filhol, who spent four months on the island examining the fauna during the French expedition to observe the transit of Venus in 1874, the Campbell Island cormorant lives on mollusca, which it scrapes off the immense patches of brown kelp that border the coasts; but no modification has taken place in the bill, which cannot be well adapted for its new use.

The same remarks apply in a measure to the tail. It is long and stiff, and is well adapted for its uses, which are: rising from the water, sitting on rocks, and probably as a rudder when diving. But in the sub-genera *Graculus* (*P. carbo,* etc.), there are fourteen tail feathers, while all the others have only twelve. This difference cannot be considered as adaptive, and could not have been accumulated by natural selection.

Probably the conclusions arrived at here will not be accepted by either Neo-Darwinians or Neo-Lamarckians. They show that the study of even a single group, like the cormorants, reveals several characters that cannot be explained by natural selection, organic selection, or use-inheritance, as these agencies preserve useful characters only; and considerable doubt has been thrown on sexual selection. It is too often the custom now-a-days to think that we have discovered all the processes working in organic nature, and that the doctrine of utilitarianism will, in some way or other, explain everything; and, when any difficulty arises, it is attributed to our ignorance of details, not of principles. These difficulties, however, must be faced; and it is possible that a close and impartial study of specific characters will destroy our complacency, and show that there is some principle of definite variation at work which preserves non-adaptive characters.

Key to the Species.

1.	Legs and feet black.	2
	Legs and feet pink or orange.	6
2.	Wing 11 inches, or more.	3
	Wing less than 11 inches.	4
3.	Tail with 14 feathers.	P. carbo.
	Tail with 12 feathers.	P. varius.
4.	Wing 9·5 to 10·5 inches.	P. sulcirostris.
	Wing 9 inches, or less.	5
5.	Abdomen white.	P. melanoleucus.
	Abdomen black.	P. brevirostris.
6.	Abdomen black.	P. chalconotus.
	Abdomen grey.	7
	Abdomen white.	8
7.	A white stripe on each side of the neck.	P. punctatus.
	No white stripe on the neck.	P. featherstoni.
8.	Fore neck black.	P. campbelli.
	Fore neck white.	9
9.	Skin on the face crimson.	10
	Skin on the face dark.	11
10.	Wing 10½ inches in length.	P. onslowi.
	Wing 11¼ inches in length.	P. ranfurlyi.
11.	Wing over 12 inches in length.	P. carunculatus.
	Wing under 12 inches in length.	12
12.	Face carunculated.	P. traversi.
	Face not carunculated.	13
13.	A white dorsal patch.	P. stewarti.
	No white dorsal patch.	P. colensoi.

The Black Shag.—KAWAU.

Phalacrocorax carbo.

Above, black, glossed with blue or green, bronzy on the back. Below greenish black; the throat and a band extending to the eye, white. A white patch on the thigh, and many linear white feathers on the head and neck when in full breeding plumage. Skin round the eye, greenish brown; beneath the eye and chin, yellow. Legs and feet, black. Eye, emerald green. Length of the wing, 12.5 to 14 inches; of the tarsus, 2.3 to 2.8 inches. Tail with 14 feathers. The young are brown with the neck, breast, and abdomen white. Egg—Bluish white; length, 2.5 inches. Found in eastern North America, Europe, Asia, Africa, and Australia, as well as New Zealand and the Chatham Islands.

The cormorant, or black shag, is found not only on the sea-coast, but also on rivers and lakes, far away from the sea. It breeds also in a great variety of situations, on bare rocks, among sedges, or in high trees. The nest is formed of sticks and grass,

(*Wood's Nat. History.*)
Black Shag.

and contains three or four eggs. The cormorant perches well on trees, holding on with its great feet. It flies heavily, but swims and dives well; never, however, using its wings when in the water, but only its feet. When engaged in fishing, it may some-times be seen swimming with its head under water, no doubt for the better observation of its prey. It can remain under water for about half a minute, in which time it travels 60 or 70 yards. On coming out of the water, it generally commences to dry its feathers, stretching out its quivering wings and preening them with its bill until everything is once more to its satisfaction.

It is rather solitary in habit, except at the breeding season

The number of eggs is three or four, and the young birds remain in the nest until they are full grown.

Mr. C. Lewis has supplied the following interesting information in regard to this bird:—

"The large black shag is a bird of heavy and laborious flight, but on occasions I have seen him develop an amount of speed to which no other bird in my experience has afforded a parallel. Some years ago, hundreds of these birds frequented Lake Ellesmere, where they wrought terrible havoc among the fish in the lake and its tributary streams. Not only so, but during the shooting season they annoyed me very much by continually raising false hopes and never coming within range. So, after the close of the season, in company with a friend, I sallied forth to conquer and destroy them.

"They came from a rookery situated on a cliff about five miles from the mouth of Lake Forsyth, and for some days we posted ourselves on the edge of cliffs which they passed on their journeys to and from the lake. In time, however, they gave these cliffs too wide an offing for our purpose, and we followed them round the coast until we found the rookery, where we wreaked an ample vengeance.

"Our tally for that year amounted to 850, and in the following year four of us could account for only 150 in three days, so that we could fairly claim to have done the State some service.

"The course which these birds took under normal conditions might be said to be along the outer edge of a semicircle. At times on the homeward route they had to encounter a heavy easterly breeze. It was on such days as these that they came home at a pace which was literally terrific. On arriving at the hill at the mouth of Lake Forsyth, they began to sail round and round in a spiral fashion, ever mounting higher and higher until they attained an altitude sufficient for their purpose; then, half crossing their wings, they simply fell the two miles or so to their home. The weight of their bodies, the poise they assumed, the height from which they descended, and the distance which they had to travel, combined to produce a rate of speed which was almost incredible. At intervals of a second or two they changed their position, leaning slightly to right or left alternately. Brushing the top of the last spur, they descended to within a few inches of the surface of the sea, whence, bending upwards in a graceful curve, they landed on their nests with but a single flap of their wings. I have seen, and shot, a cock pheasant floating down a gully; but I never felt like beginning to get anywhere near one of these black shags. A crossing shot conveyed no hope of getting within a chain of its object, and if you trained the gun upon a black speck in the distance, proposing to let go at an incoming bird, the swerves I have mentioned hopelessly bankrupted all chance of success. And we all wanted to kill one, for its fall would have been worth seeing.

"The noise occasioned by the rushing of their pinions in their swerving flight was considerable, and whenever I heard the roaring sound which heralded their approach, I could not help recalling: 'And suddenly there came a sound from heaven, as of a rushing mighty wind.'"

The Pied Shag.—KAWAU.

Phalacrocorax varius.

Above, greenish black; brown on the back. Below, white; the thighs with a greenish black patch. Skin on the side of the face, light blue. A spot in front of the eye, bright yellow. Legs and feet, black. Eye, seagreen. Length of the wing, 11.75 ; of the tarsus, 5.5 in. The young birds are brown above; and white mottled with brown below. Eggs—Pale blue; length, 2.4 in. New Zealand only.

The pied shag is the common coast shag of the North Island, but it is not so common in the south. It is, however, by no means confined to the coast, as it may be found far away in the interior. Like the black shag, it builds in various localities, and lays from two to four eggs. It is not so gregarious as most of the shags, and a single bird may often be seen sitting on a jutting-out rock, or on a branch of a tree overhanging the water.

The Little Black Shag.

Phalacrocorax sulcirostris.

Brownish black, with a dull oil-green gloss. The upper back and wing coverts, grey; each feather with a black border. In the breeding season there are a number of white plumules on the head and neck. Skin on the sides of the face, brownish black. Legs and feet, black. Eye, deep green. Length of the wing, about 10in: of the tarsus, 1.75 in.; of the bill, 1.75 in. Java to Australia, New Caledonia, and Norfolk Island.

In New Zealand a single colony of little black shags is known at the Bay of Islands; probably this bird is a late arrival from Norfolk Island.

(Buller.)

Pitt Island Shag.   Frilled Shag.

The Frilled Shag.

Phalacrocorax melanoleucus.

Above black; below white. The feathers on each side of the neck lengthened in the breeding season. Thighs with a black patch. Bill, yellowish. Skin on the face, dull olive; that on the chin, yellow. Legs and feet, black. Eye, dark brown. Length of the wing, 9 in.; of the tarsus, 1.45 in.; of the bill, 1.2 in. Length of the egg, 1.85 in. The young bird has the abdomen mottled with black. Australia, New Guinea, and New Caledonia.

In New Zealand this bird is rare, but it is occasionally found in both islands. In the Canterbury Museum there are specimens shot at Sumner in 1886.

Nest of White-throated Shag.

The White-throated Shag.—AROAROTEA.

Phalacrocorax brevirostris.

Black, with the throat and chin white. Bill, yellow. Skin on the face, greenish yellow, that on the chin, yellow. Legs and feet, black. Eye, chocolate brown. Length of the wing, 9 in.; of the tarsus, 1.4 in.; of the bill, 1.1 in. The young are entirely black. Egg—Bluish white; length, 1.9 in. The variable colouring of this bird, and how it often approaches the last species, have already been mentioned. Probably the best distinguishing marks between the two are the ear coverts, which are black in *brevirostris* and white in *melanoleucus*. New Zealand only, where it is common in both Islands.

The white-throated shag is not a coast bird, but is found chiefly near the end of long winding sounds and inlets, or else on the rivers. It not uncommonly forms rookeries with the black shag. It feeds not only on fish, but also on fresh water shrimps. It is a sociable bird, congregating in considerable numbers, especially for breeding. The nest is large and is made of sticks lined with grass. The eggs are four in number.

(*Buller.*)

Spotted Shag.   Chatham Island Shag.

The Spotted Shag.—PAREKAREKA.

Phalacrocorax punctatus.

Above, grey, with black spots. Tail and thighs, black. Below, lead grey. Head and neck, black, with a white stripe down each side of the neck. Head, in the breeding season, with two crests. Skin on the sides of the face, dark blue. Legs and feet, yellow. Eye green. (Mr. E. J. Haynes, taxidermist, states that the eye is very dark hazel, but becomes green after death.) Length of the wing, 9.3 in.; of the tarsus,

2.25 in. In the young the upper surface is grey with dark spots, the lower surface, greyish white; and the skin on the face is yellowish flesh colour; the feet and legs dull flesh colour. Egg—White; length, 2.4 in. New Zealand only.

This is the common coast shag of Canterbury and Westland. It is rare in the north, but is found on the eastern side of Hauraki Gulf. It never goes inland, but breeds on sea cliffs. The breeding season is from the middle of October to early summer, and the bird lays three eggs. Both sexes sit on the eggs. The young are born blind and almost naked, but soon become covered with down, which is dark brown above and whitish below. None of the ocean shags goes far from land, and they never dive from the air, but always first settle on the water. They swim low, the tail being about level with the surface. They rise on the wing from the water with difficulty, having to make three or four leaps with the body nearly upright. When changing the feeding ground, they fly in companies, usually at a low elevation, keeping just above the curl of the wave. They follow the shoals of fish, and their great power of diving ensures a plentiful supply of food. After feeding they repair to the cliffs and bask in the sun.

The Chatham Island Shag.

Phalacrocorax featherstoni.

Upper surface, dark greenish grey, with black spots. The head and throat, nearly black. Breast and abdomen, silvery grey. Head, in the breeding season, with two crests. Skin on the face, purple. Legs and feet, orange. Eye, pale green. Length of the wing, 9 in.; of the tarsus, 1.9 in. Egg—White; length, 2.4 in. Chatham Islands.

The Campbell Island Shag.

Phalacrocorax campbelli.

Upper surface and thighs blue black; the shoulders and wing-coverts green black. A narrow white bar on the wing. Whole under surface, except the fore neck, which is black, white. Head crested. Skin on the face dark blue, with minute crimson spots; chin bright orange; legs and feet flesh-colour; eye dark brown. Length of the wing, 10.5 in.; of the tarsus, 2.4 in. The young has the upper surface and fore neck brown, glossed with greenish; the chin, throat, and abdomen white. There is no white band on the wing. Skin on the face dull orange, with crimson spots; the chin orange. Campbell and Auckland Islands.

The Auckland Island Shag.

Phalacrocorax colensoi.

Like the last, but with a narrow band of white down the fore neck. Skin on the face purple, the chin crimson, legs and feet flesh colour, eye dark brown. Length of the wing, 10.7 in.; of the tarsus, 2.4 in. In the young, the front of the neck is dark for a short distance as in *P. campbelli.* Auckland Islands. Intermediate varieties between this species and the last are not uncommon, and it seems possible that they may interbreed.

(*Voy. Erebus and Terror.*)
Pink footed   Shag.

The Pink-footed Shag.

Phalacrocorax chalconotus.

Entirely greenish black, with flesh-coloured legs and feet. Skin on the face black, the eyelids blue; chin orange. Head crested in the breeding season. Eye brown. Length of the wing, 11.2 in.; of the tarsus, 2.5 in. Egg—White; length, 2.7 in. Stewart Island and the southern coasts of Otago.

The Stewart Island Shag.

Phalacrocorax stewarti.

Above black, glossed with bluish green; below white. A white band on the wing, and a white patch on the lower back. Skin on the face black, the eyelids blue, chin orange, legs and feet flesh-colour. Head crested in the breeding season. Eye brown. Length of the wing, 11.6 in.; of the tarsus, 2.8 in. Stewart Island and the southern coasts of Otago. There is a specimen in the Otago University Museum with well marked red caruncles, but usually these are absent.

Stewart Island Shag.

The Pitt Island Shag.

Phalacrocorax onslowi.

Colours like the last, but the white patch on the back is sometimes wanting. Skin on the face orange, the eyelids blue; some yellow caruncles at the base of the bill. Legs and feet orange. Head crested. Eye brown. Length of the wing, 10.4 in.; of the tarsus, 2.5 in. Egg—White; length, 2.45 in. Pitt Island, Chatham Group. See page 312.

The Bounty Island Shag.

Phalacrocorax ranfurlyi.

Colours like the last, but no dorsal white patch, and no caruncles. Skin on the face crimson with black dots, the eyelids purple; legs and feet flesh colour; eye brown. Length of the wing, 11.25 in.; of the tarsus, 2.25 in. The birds are crested in August. Young—with the upper surface, the sides of the head, and the neck greyish brown. Chin, breast, and abdomen white. Skin on the face brown, the eyelids purple; chin pouch grey. Legs and feet pale flesh-colour. Bounty Islands.

The Rough-faced Shag.

Phalacrocorax carunculatus.

Colours like the last, but a wide dorsal patch always present, and the nasal caruncles well developed. Skin on the face chocolate brown, the eyelids bright blue; the caruncles orange. Legs and feet pale flesh-colour. Usually there is no trace of a crest, but occasionally a vestigial one occurs. Eye brown. Length of the wing, 12.3 in.; of the tarsus, 2.6 in. Egg—White; length, 2.5 in. Breeds at White Rocks in Pelorus Sound. Eggs and young in the middle of July. *P. huttoni*, of Buller, is the young of this species.

The Macquarie Island Shag.

Phalacrocorax traversi.

Upper surface black; bluish on the head, neck, and lower back; greenish brown on the back and wings. A narrow white alar bar. Lower surface white, the white extending over more than half the neck. A patch of bluish black on the thighs. Skin on the face dark blue, with numerous red or orange caruncles behind the nostrils; chin pouch dark, spotted with orange. Legs and feet flesh-colour. A small crest on the head. Length of the wing, 12 in.; of the tarsus, 1.7 in.; of the bill, 2 in. Macquarie Island.

ORDER PYGOPODES.

Bill long and straight. Tail very short. Legs flattened, set far back.

Genus Podicipes.

Wings short and pointed. Toes long, margined on the sides and united at the base to the middle toe; hind toe strong and strongly lobed. All parts of the world.

Crested Grebe.   (*Meyer.*)

The Crested Grebe.—KAHA.

Podicipes cristatus.

Above brown, below white; top of the head black, with a double crest. Chin white; ruff on the neck rufous, tipped with black. Shoulders and a band on the wings white. Eye crimson. Length of the wing, 7.75 in.; of the tarsus, 2.6 in. The neck and breast are occasionally tinted with rufous brown. Egg—At first greenish white, then yellow brown; length, 2.35 in. Europe, Asia, Africa, and Australasia.

Formerly this bird was often seen on the larger lakes of the South Island, but the tourist traffic has disturbed it. It still haunts these smaller and less frequented lakes, and occasionally makes an excursion into inhabited districts. In Europe it is migratory, but it does not appear to be so in New Zealand. It is an excellent diver, and feeds chiefly on aquatic insects.

According to Mr. Potts, it swims low in the water with an air of demure gravity, which affords a marked contrast to the rapid movements of most of the other natatorial birds, with which it frequently associates. He found the nest in November and December. The structure is large and very solidly built of pieces of decayed *Carex virgata* raised about a foot above the surface of the water, its sloping sides giving a ready means of reaching the basin-like depression on the top, in which the eggs are deposited.

Mr. W. T. L. Travers says that this bird was found at all seasons of the year upon Lake Guyon, a small lake in the Nelson province, lying close under the Spencer Mountain range. The water of this lake is generally very warm, and even in severe seasons has never been frozen over. To this fact is attributed the circumstance that some of these birds are found upon it throughout the year. There are several apparently permanent nests on the borders of the lake, which have been occupied by pairs of birds for many years in succession. Therefore, Mr. Travers infers that these birds pair for life. The nests, he says, are built amongst the twiggy branches of trees which have fallen from the banks of the lake, and now lie half floating in the water, and are formed of irregularly laid masses of various species of pond weeds found growing in the lake, which the birds obtain by diving. The nests are but little raised above the surface of the water, for, in consequence of the position and structure of its feet, and the general form of its body, the grebe is unable to raise itself upon the former unless the body be in great measure supported by water.

The eggs are usually three in number. Both male and female assist in the labour of incubation, although it is supposed that the chief part of this task devolves upon the female, and that she is relieved by her partner only for the purpose of enabling her to feed.

The Little Grebe.—WEWEIA.

Podicipes rufipectus.

Above, blackish brown, finely streaked with white on the head. The throat, brown; breast, rufous; and the abdomen, white. Breast sometimes clouded with dirty yellow. Eye, silvery grey. Length of the wing, 4.75 in.; of the tarsus, 1.4 in. Egg—White; length, 1.65 in. New Zealand only.

The little grebe, or dabchick, is fairly common in suitable localities in both Islands; but it is wary and is seldom seen. It is necessary to lie in wait, silent and patient, and then it may be seen swimming gaily about, but never going far from the sheltering rushes. On the least alarm it dives and is seen no more.

Little Grebe.

(*Voy. Erebus and Terror.*)

The nest is made among the rushes, floating on the surface of the water. It is a rather large and somewhat clumsy structure, formed of the roots and leaves of various aquatic plants. Mr. Potts has found a nest built against the stem of *Carex virgata,* beneath the drooping leaves of which it was perfectly concealed from casual observation. Situated just within the swampy side of a small lake, it was raised a few inches only above the water level.

ORDER LAMELLIROSTRES.

Bill straight with a distinct nail at the tip of the upper mandible, and with a series of thin plates on each side. Tarsus, moderate, reticulated behind and generally in front; the anterior toes fully webbed.

Key to the Families.

1.	Hind toe narrowly lobed.	Anatidæ.
	Hind toe broadly lobed.	2
2.	Bill decidedly depressed.	Fuligulidæ.
	Bill more or less compressed.	3
3.	Edges of mandible not serrated.	Merganettidæ.
	Edges of mandible serrated.	Mergidæ.

Family Anatidae.

Bill flat and broad. Wing with metallic speculum. Hind toe very narrowly lobed.

Key to the Genera.

1.	Outer webs of the tertials chestnut.	Casarca.
	Outer webs of tertials not chestnut.	2
2.	Bill much wider at the end than at base.	Spatula.
	Bill not much wider at end than at base.	3
3.	Bill as long as the head.	Anas.
	Bill shorter than the head.	4
4.	Bill without membrane near the tip.	Nettion.
	Bill with a narrow membrane near the tip.	5
5.	Wings well developed.	Elasmonetta.
	Wings short.	Nesonetta.

In April, 1903, the *Akaroa Mail* published an interesting article dealing with some members of this order on Banks Peninsula. "Our native water fowl, unlike those of Britain," the *Mail* said, "are not migratory, but live with us all the year round. The grey and paradise ducks, spoonbill ducks, and black and brown teal, all used to go inland to breed. The young were reared in the creeks, running in every part of the Peninsula, and, in the autumn, when they grew strong enough on the wing, the parents led them to Lakes Ellesmere and Forsyth, and other lagoons and marshes bordering on the sea, for it is in these salt marshes that these birds get the food they love, and grow to that plump perfection which renders them so delicious. When the bush covered the Peninsula, the creeks were well sheltered and little disturbed. In those days vast numbers of the birds were

reared, and the clouds of ducks that used to frequent the country
from the head of Lake Forsyth along Birdling's Flat and Lake
Ellesmere to Taumutu were enormous.'' Twenty years ago, it is
also stated, even a good many grey ducks were reared in the
Akaroa creeks, and ten years later a brace or two still came, but
these breeding-places are now so often disturbed, and there are
so many bad sportsmen about who cannot resist shooting at a
wild duck at any season of the year, that the tribute of young
ducks from inland parts of the Peninsula to the coast has almost
ceased. ''In Gough's Bay and other localities, where good sportsmen live,'' continues the *Mail*, ''a few are still reared in peace;
but, where thirty years ago thousands reached maturity, only a
few now survive to attend the great autumn muster.''

Genus Casarca.

Bill as long as the head, as wide as high at the base; not
broader at the tip than at the base. Second quill the longest.
Toes long, fully webbed; the hind toe elevated, lobed. The
Eastern Hemisphere.

The Paradise Duck.—PUTANGITANGI.

Casarca variegata.

Male—Head, neck, and breast, black; back, black, pencilled transversely with white. Abdomen, ferruginous, pencilled with black. Wing
coverts, white. Eye, black. Female—Head and neck, white; and the
breast, ferruginous, like the abdomen. Length of the wing, 14.5 inches; of
the tarsus, 2.5 inches. Egg—Pale cream colour; length, 2.75 inches. New
Zealand only.

This bird is common in the South Island, but rare in the
North Island. It is remarkable that the female, with its white
head, should be more conspicuous than the male, notwithstanding
that it sits upon the eggs. The young, however, resemble the
male, which is another important peculiarity Their usual
breeding-place is in the wide river-beds of the South, but they
sometimes build in trees. The old birds are as good adepts at

stratagems for luring the wayfarer from their nests or young as any birds in Europe.

Mr. Travers states that this duck is usually found in the valleys, feeding more upon the tender shoots of young grass and upon herbs of different kinds than upon fish or other forms of animal life. It breeds from October to January, and not infrequently rears two broods in one season. Both parents are anxious and watchful about their young, resorting to the ruse of pretending

Paradise Duck: male.   *(Voy. Erebus and Terror)*

lameness, and inability to rise from the ground, in order to draw off any animal which they think likely to be mischievous. Upon the danger signal being uttered by the parent bird, the young ones usually make at once for the nearest flowing water, down which they float close to the bank, seeking cover, and availing themselves with great sagacity of every opportunity of shelter or concealment, in which they are assisted by their similarity in general colour to the soil and vegetation.

Grey Duck.

Genus Anas.

Bill broad, as long or longer than the head; higher than broad at the base; of nearly equal breadth throughout. Hind toe small, narrowly webbed. Cosmopolitan.

The Grey Duck.—PARERA.

Anas superciliosa.

Greyish brown varied with yellowish white. Eyebrows, cheeks, and upper part of the neck, yellowish white, with two small bands of blackish

brown on the cheeks. Speculum, green, margined above and below with
black. Eye, bright hazel. Length of the wing, 10.5; of the tarsus, 1.65
inches. Egg—Cream colour; length, 2.5 inches. From Java, through
Australia and Polynesia to New Zealand and the Chathams, Auckland,
and Campbell Islands.

The nest of the grey duck has been found in many situations.
It has been seen by Mr. Potts close by the edge of a bush creek,
amongst damp ferny shades; out on the plain, sheltered by a
tussock, quite far away from water; and often on a hillside;

Grey Teal.   (*Pro. Zool. Society.*)

sometimes even in trees. But wherever the cup-shaped nest is
found, whether on the level plain or in a swamp, it is profusely
lined with down, and diffuses a strong musky odour. The grey
duck, fortunately, still maintains its ground, in spite of heavy
losses each year in the shooting season. In the autumn, when the
harvest is over, it resorts in the evening to feed in the stubble
fields, and is specially partial to pea stubbles.

Genus Nettion.

Bill rather compressed and shorter than the head. Central tail feathers more or less acuminate, and extending somewhat beyond the lateral ones. Cosmopolitan.

The Grey Teal.—TETE.

Nettion castaneum.

Greyish brown, the lower surface spotted with dark brown. Sides of the head, brownish white. Greater wing coverts, white; the speculum, black, tipped with white. Bill, legs, and feet, bluish black. Eye, yellowish brown. Length of the wing, 8 inches; of the tarsus, 1.3 inches. Egg— Cream colour; length, 1.95 inch. Java to Australia and New Zealand.

The grey teal breeds in the North Island, but seems to be only an occasional visitant to the South, where specimens have been obtained even in Otago. Very little is known about its habits.

Genus Elasmonetta.

Bill compressed; soft on the edges of the apical part of the upper mandible; lamellæ extremely developed. Wings well developed. New Zealand only.

The Brown Duck.—PATEKE.

Elasmonetta chlorotis.

Above greyish brown, varied with rufous; breast, rufous; abdomen, yellowish brown spotted with black. Speculum, greenish black, bordered above and below with rufous white. In the male, the head is tinged with green, and usually there is a white band on the fore part of the neck. Eye, black. Length of the wing, 8 inches; of the tarsus, 1.5 inches. Egg— Cream colour; length, 2.25 inches.

According to Sir W. Buller, it is a poor flier, but dives well. The nest is made of grass, thickly lined with down, sometimes close to the edge of a swampy creek, or beneath the shelter of a large Maori head (*Carex virgata*).

The brown duck inhabits the Chatham Islands as well as New Zealand. Here it was formerly abundant, but it has become very rare in the south. It frequents small streams and ponds, and is not seen on lakes.

(*Voy. Erebus and Terror.*)

Brown Duck.

Genus Nesonetta.

Like the last, but the wings shortened so that the bird cannot fly, and the lamellæ of the bill less developed. Tail wedge-shaped in the adult. Claws strong. Auckland Islands only.

The Flightless Duck.

Nesonetta aucklandica.

Above blackish brown, some of the feathers with green reflections and margined with rufous. Cheeks and sides of the neck speckled with dusky white; breast ferruginous, varied with obscure spots. Abdomen brown, banded with dusky white. Eye, dark hazel. Length of the wing, in the male 5.5 in.; in the female 5 in.; of the bill, in the male 1.6 in.; in the female 1.35 in.; of the tarsus, 1.25 in. Auckland Islands.

The flightless duck is not rare in the Auckland Islands, where it lives on the sea-coast, running over the masses of brown seaweed or kelp, which abounds on the coasts, and climbing over the slippery rocks by the help of its strong and sharp claws. When pursued, it never attempts to fly, nor does it dive, but scurries away over the rocks near the sea-shore. Some members of this

Flightless Duck.　　*(Voy. Erebus and Terror.)*

species were placed on the Kapiti Island Sanctuary by Dr. L. Cockayne in December 1907. Early in 1908, they were reported to be getting on very well.

Genus Spatula.

Bill longer than the head, much dilated for half its length from the tip; the lamellæ long. Cosmopolitan.

The Shoveller.—KURUWHENGI.

Spatula rhynchotis.

Male—Above dark brown, the head and neck grey, with a white line near the bill. Breast brown, varied with rufous white. Abdomen dark rufous. Wing-coverts and a longitudinal stripe on the wings ash-grey. Eye, bright yellow. Female—Brown. Length of the wing. 9.8 in.; of the tarsus, 1.3 in. Egg—Greenish cream-colour; length 2.1 inch. Australia, Tasmania, New Zealand, and the Chatham Islands.

The shoveller, or spoonbill duck, is common in both islands, and frequents the muddy shores of lakes and streams. It feeds on small aquatic insects, which its large bill is well adapted to catch. It is essentially a surface feeder, and never dives for its food.

According to Mr. Gould, the New Zealand male bird has more white on the sides of the lower neck and breast than is seen in those from Australia.

Mr. Potts says that this bird is very evenly distributed over the country, although large numbers are to be met with only in certain localities in the North Island, where it is of common occurrence. For some years after settling in Canterbury, he regarded the bird as quite a *rara avis*, but later on it was seen much oftener than formerly. The nest is made of fine grass, lined with a small quantity of down, and is sometimes placed on the side of a hill far away from water.

Family Fuligulidae.

Bill depressed, the hind toe broadly lobed.

Key to the Genera.

Bill not wider at the tip.	Nyroca.
Bill rather wider at the tip.	Fuligula.

Genus Nyroca.

Bill not distinctly wider near the tip than at the base; neither very broad nor short. Cosmopolitan.

The White-winged Duck.—KARAKAHIA.

Nyroca australis.

Head and neck dark reddish brown, back and abdomen brown, lower breast and under tail-coverts white, wing feathers white, tipped with brown; the speculum white. Bill black, with a slate coloured band near the tip. Legs and toes grey, the webs black bordered with grey. Eye, white. Length of the wing, 8.5 in.; of the tarsus, 1.5 in. New Guinea, Australia, Tasmania, New Caledonia, and New Zealand.

The white-winged duck breeds round the margins of the lakes of the lower Waikato. Formerly, it was very abundant on Lake Rotomahana, but almost the whole flock seems to have been

(*From a Drawing in the Ch. Ch. Mus.*)
White-winged Duck.

destroyed by the eruption of Mount Tarawera in 1886. It sits all day on the water in the middle of the lake, and is very difficult to approach. The white on the wing is very conspicuous when flying. It is an excellent diver, and gets much of its food below the surface of the water.

Genus *Fuligula*.

Bill somewhat broader near the tip than at the base, rather broad and short and rounded at the tip. Northern Hemisphere and New Zealand.

The Black Teal.—Papango.

Fuligula novae-zealandiae.

Male—Head and neck black, glossy with purple above and green on the sides. Upper surface and breast black, abdomen brownish white, speculum white. Bill blue, tipped with black; legs and toes grey, the webs black margined with grey; eye bright yellow. Female—Above dull black; below brown, mottled with white; a band of white round the upper mandible. Length of the wing, 7.75 in.; of the tarsus, 1.25 in. Egg—Greyish cream colour; length, 2.1 in. Both islands of New Zealand, but more common in the North.

(Voy. Erebus and Terror.)
Black Teal.

This species, in colour, is very like the scaup duck of Europe, but, unlike that species, it does not frequent the sea-shore, but lives among rushy streams in small flocks.

In the hill country of the South Island, about 1870, it was fairly common. It is a gregarious bird, and it delights to assemble in flocks. It may be seen, on some of the more secluded lakes, swimming about and disporting with numbers of other water-fowl, very frequently diving. Sometimes it breeds in the shelter of a huge nigger-head. Mr. Potts has found its nest well concealed by a large snow-grass tussock, within a few feet of water, where there was a rent or crack in the ground. The nest is of grass, thickly lined with down, with five eggs.

According to Mr. Travers, those that occupy Lake Guyon are generally to be found sitting on half-submerged logs close to the bank, from which they appear to watch the small fish, upon which they chiefly feed. Like the other birds on the lake, they are by no means shy, quietly dropping into the water and swimming away if approached too closely. Formerly, the black teal was very abundant in the Lower Waikato district.

Family Merganettidae.

Bill more or less compressed, the tail rather long and stiff. No tooth-like serrations on the edges of the mandibles. Waigiou, New Zealand, and the Andes.

Genus Hymenolaemus.

Bill as long as the head, furnished with a soft membrane for half its length from the tip. Wings short, with large callosities on the joints. No speculum. Hind toe strongly lobed. New Zealand only.

The Blue Duck.—WHIO.

Hymenolaemus malacorhynchus.

Above lead blue; below the same, spotted with rufous and varied with white. Bill pinkish white, legs and feet dark brown, eye bright yellow. Length of the wing, 9 in.; of the tarsus, 1.8 in. Egg—Cream colour; length, 2.7 in. Both islands, and formerly, it is reported by Captain Bollons of the "Hinemoa," in the Auckland Islands.

It was the opinion of Mr. Potts that the only way of seeing this singular bird to advantage was by paying a visit to the mountainous districts. "On a mountain torrent, where the foaming water dashes from rock to rock in countless eddies," he writes, "the blue duck lives at ease, making its way up or down stream. Sometimes it may be observed basking in the sunshine, near a shallow pool of the rapid streamlet. Sometimes it is a burrower, and its nest may be found in a hole in a bank." He has found it concealed from view by overhanging sprays of those various alpine veronicas which sometimes make the mountain creeks gems of beauty. The nest, like that of other ducks, is thickly lined with down, and generally contains five eggs. "When the parent birds have their brood in charge," he adds, "they certainly exhibit much less craft, as well as energy, for the protection of their offspring, than any other duck. With them there is little, if any, attempt at concealment of the young; none of the ruses which, with the paradise duck, often prove successful in misleading their enemies, are brought into requisition. Usually they simply drop down the rapid, trusting for escape, apparently, to the turbulence of the stream, which is an asylum safe enough

(*Gray's Genera of Birds.*)

Blue Duck.

from most, if not all, indigenous persecutors, but not from the settler's dog. They seem loth to land, and, if compelled to do so, their progress is not very rapid; in fact, they impress one with an idea of their helplessness. The duck marches in front, with her low wailing call, the small brood follow while the drake protects the rear, or rather offers himself as the first victim to the pursuer. In winter, they congregate in flocks of moderate numbers."

The blue duck's peculiar shrill and sibilant note, which is sometimes distinctly heard over the noise of the loudest cataract, has been commented upon. Mr. Travers says that these birds appear to have been especially endowed with this note in consequence of their frequenting certain localities. He found them on Lake Guyon very tame, looking with an appearance of surprise, mixed with a dash of stupidity, at intruders on their privacy, and rarely taking to the wing unless closely pursued, and then only flying to a short distance. They breed in November and December, and, like the paradise duck, sometimes bring up two broods in the year. In seeking for food, they usually stand on a stone in the middle of some rapid, from which they pick up any stray article of diet which is being carried by; whilst they are also constantly seen busily engaged in searching for food under the water in the rapids. In doing this, they use their wings like hands to cling to the stones in order to assist them in overcoming the rush of the water. They appear to be much attached to their young, but use no stratagem to draw off an enemy, whilst the young merely move from spot to spot to escape danger, rarely diving.

Family Mergidae.

Bill compressed, without any lamellæ on the sides, but with a series of distinct tooth-like serrations on the edges of both mandibles.

Genus Merganser.

Serrations of the bill very conspicuous, and inclined backwards at the tips. Northern Hemisphere and Brazil.

The Southern Merganser.

Merganser australis.

Head and neck brown, with a rufous tinge on the throat and lower neck; rest of the body dark grey, with grey and white transverse markings on the breast. Lower abdomen and tail-coverts whitish. Speculum white. Bill dark olive, the base of the lower mandible orange. Legs and feet orange, the webs dusky. Eye almost black. The head is more or less crested. The male has more white on the wing, the abdomen is paler, and the brown on the throat runs further down. In the young, the upper surface is dark olive-brown; the throat, fore neck, and a spot under the eye bright rufous. Under surface yellowish white. Length of the wing, 7.25 in.; of the tarsus, 1.4 in. Auckland Islands.

Southern Merganser.

(Voy. au Pole Sud.)

The southern merganser is our only sea duck, and it does not frequent the coast and open waters, but only the sheltered harbours. Although its wings are short, it flies well. It feeds on fish, which it catches by diving. It does not appear to be much alarmed at man and his works, for when Lord Ranfurly was in the Auckland Islands, collecting birds, an old male bird flew close up to the steamer, where it was anchored for the evening, settled in the water within a few yards of the vessel, and swam calmly about quacking like a domestic duck.

How the southern merganser got to New Zealand is a puzzle. It is a northern genus, the members of which migrate southwards in the winter to the Black Sea and China; while in America these go as far as the Bermudas. There is also a species in Brazil, but none between that and New Zealand.

Sub-class Ratitae.

Breast-bone without a keel. Barbs of the feathers disconnected. Flightless.

Family Apterygidae.

Bill much longer than the head, the nostrils near the tip. Three toes in front and one behind. Wings rudimentary. New Zealand only.

Genus Apteryx.

Tarsi about the length of the middle toe, very robust; lateral toes equal; hind toe short, elevated above the others.

Key to the Species.

1. Brown with longitudinal streaks.		2
Grey with transverse bars.		3
2. Feathers soft to the touch.	A.	australis.
Feathers of the back harsh to the touch.	A.	mantelli.
3. Smaller, bill straight.		4
Larger, bill curved.	A.	haasti.
4. Bars confused.	A.	oweni.
Bars distinct.	A.	occidentalis.

In the absence of the extinct moa, the kiwi is the most notable living bird of New Zealand, and should be classed among the dominion's treasured possessions.

It is an anomaly. While it has been aptly described by Dr. Wallace as one of the queerest and most unbird-like of living birds, it apparently represents several widely different orders in its heterogeneous structure. Sir R. Owen says that it seems to have borrowed its head from the long-billed waders, its legs from the gallinæ, which include the domestic fowls and other scratchers, and its wings from the struthious birds, which include the ostrich, the rhea, the cassowary, and the emu.

There is very little that we do not know about the kiwi. Not a bone in its frame, or a muscle on its ungainly body, and hardly a feather in its hair-like plumage, has escaped minute and elaborate description. Its innermost private life has been invaded; and its habits, its clumsy gait, its wretchedly defenceless condition, its family failings, its deformities and malformations have been made public. The only thing that is hidden from us is its origin, which is still a deep mystery, though it is certain that the members of the family Apterygidæ, which belong to the struthious birds, have a very ancient lineage. Kiwis have a more generalised structure than that of other struthious birds. They therefore belong to a separate type, and cannot with any degree of correctness be said to represent the extinct moas.

At first glance, the kiwi's body seems to be covered with hairs instead of feathers; outwardly, it has neither wings nor a tail; and the position of its nostrils at the tip of its long and slender beak, instead of at the base, constitutes one of the most distinguishing features. Its olfactory organs are remarkably perfect, and it has a keener sense of smell than is possessed by any other living bird. Its eyes, however, are small and inefficient, differing remarkably from those of other birds by the absence of a vascular membrane called the pecten. Professor T. J. Parker, however, states that this peculiarity applies in strictness only to the adult, and that in advanced embryos a small but distinct pecten is present. The female is considerably larger than the male, and the largeness of the egg she lays is out of all proportion to the size of the bird. When the trunk is stripped of its plumage, Sir R. Owen remarks, the body presents the form of an elongated cone gradually tapering forwards, from the broad base formed by the haunches to the extremity of the attenuated beak; and the wings appear as two small crooked appendages, projecting about an inch and a half from the sides of the thorax, and terminated by a curved, obtuse, horny claw, a quarter of an inch long.

Professor Parker, in his *History of the Kiwi*, deals with evidence which seems to him to indicate that the ancestors of Apteryx had the interrupted pterylosis, or feather-arrangement, characteristic of the Carinatæ, and that once upon a time their remarkable fore-limbs were true wings, which have been lost, probably for want of usage. A minor matter which, to his mind, points to the same conclusion, is the fact that a sleeping kiwi assumes precisely the same attitude as an ordinary carinate bird, the head being thrust under the side feathers, between the body and the upwardly directed elbow. "On the whole," he says, "it will be seen that the study of the development of the kiwi tends to lessen the gulf between it and ordinary birds, and to show that its ancestors probably possessed many of the more important and distinctive features which characterise the Carinatæ of to-day. The facts clearly indicate that the founder of the apterygian house had interrupted plumage, functional wings, an ordinary avian tail, a keeled sternum, a double-headed quadrate, lateral optic lobes, and a pecten in the eye; in other words, that the ancestors of the genus were typical flying birds, and not bird-like reptiles."

As to the relation of the kiwi to the other genera, Professor Parker finds that it has been shown to be most nearly allied, as far as its skeleton is concerned, to the moa, differing from it, however, in many important respects. He says that it must certainly have been isolated at a very distant period, and, as far as can be ascertained, some of its more striking peculiarities are distinctly correlated to its method of feeding. "Most nocturnal animals have large eyes, suited for taking the utmost advantage of the semi-darkness," he concludes, "but the kiwi, finding its prey by scent alone, has developed an extraordinarily perfect olfactory sense, while, at the same time, having no need to keep watch against beasts of prey, its eyes have diminished in size and efficiency to a degree elsewhere unknown in the bird class."

The kiwi's mode of reproduction caused a great deal of controversy before its habits became well known. Sir George Grey, in 1863, forwarded to Dr. Sclater, in England, a letter he had received from a gentleman in Hokianga, who said:—"Several years ago an old native, who had been a kiwi hunter in the times when kiwis were plentiful, told me a strange tale about the manner in which the kiwi hatches its eggs. I, of course, cannot vouch for the correctness of the story, but think it worth relating. He said that the kiwi did not sit, like other birds, upon the egg, but under it, first burying the egg in the ground to a considerable depth, and then digging a cave or nest under it by which about one-third of the lower end was exposed, and so lying under the egg and in contact with the lower end, which came, as it were, through the roof of the nest or burrow. The appearance of the egg, which I propose to send, corroborated the statement, for two-thirds of its length, the small end, was perfectly clean and white, and about one-third, the large end, was very much discoloured and very greasy, evidently from contact with the body of the bird. The difference in colour and condition of the ends of the egg was quite remarkable, and well defined by a circular line passing round the egg.''

Mr. A. D. Bartlett, Superintendent of the Zoological Society's Gardens in London, endeavoured to put these statements to a test by means of two kiwis. They showed a desire to pair by the loud calling of the male, which was answered by the female in a much lower and shorter note. They were particularly noisy at night, but were quite silent in the day time. The female laid two or three eggs, but as soon as she quitted the nest, the male bird took to it, and remained constantly sitting. By-and-by, the birds occupied the two opposite corners of the room in which they were kept, the male being on the two eggs in the nest under the straw, while the female was concealed in her corner, also under a bundle of straw placed against the wall.

During the time of incubation they ceased to call at night; they were perfectly silent, and remained apart. The eggs were found in a hollow formed on the ground in the earth and straw, and placed lengthwise side by side. The male bird lay across them, his narrow body appearing not sufficiently broad to cover them in any other way. The ends of the eggs could be seen projecting from the side of the bird. He continued to sit in the most persevering manner until he was exhausted, and he then left the nest. On examining the eggs, no traces of young birds could be found, but the superintendent says that, notwithstanding this failure, there was sufficient to show that the kiwi's mode of reproduction does not differ essentially from that of the allied struthious birds, as in all cases that have come under his notice only the male bird sits. "I have witnessed the breeding of the mooruk, the cassowary, the emu, and the rhea," he says, "and the mode of proceeding of the Apteryx fully justifies me in believing the habits of this bird to be in no way materially different from those of its allies."

The Rev. J. G. Wood, in his *Natural History of Birds,* says that the eggs laid at the London Zoological Gardens by *Apteryx australis* are indeed wonderful, for the bird weighs just a little more than four pounds, and each egg weighs fourteen or fifteen ounces, its length being 4¾ in., and its width rather more than 2 in. An egg of Owen's Apteryx in the Canterbury Museum, obtained from the West Coast, measures 4½ in. in length, with a breadth of 2 7-12 in. The egg of an *Apteryx mantelli* has measured over 5 in. in length, and 3 in. in breadth.

There are four or five species of kiwis. One belongs to the northern portion of the dominion, and the others to the south. All possess the same marked peculiarities, but their habits may be described as observed in the various species. Mr. Potts's articles in the *Transactions of the New Zealand Institute,* which could hardly be improved upon, are the chief source from which information has been drawn.

The Brown Kiwi.—KIWI.

Apteryx mantelli.

Rufous brown, streaked longitudinally with black. Feathers of the back harsh to the touch, owing to the shafts being prolonged beyond the barbs. Bill slightly curved. Claw on the wing much curved, and black. Eye black. Male—Length of the bill, 3.75 in. to 4.25 in.; of the tarsus, 2.5 to 2.75 in. In the female—Bill, 5.2 in to ¾ in.; of the tarsus, 2.75 in. to 3.5 in. Egg—White; length, 4.75 in. North Island and Little Barrier Island.

The northern species, the kiwi-nui of the Maoris, is now rare. Its plumage is darkest on the back, and the feathers on its back differ from those of other species by being harsh to the touch.

It inhabits the darkest and densest forests. In the early days of colonisation, members of the species must have been numerous, for the late Mr. Allan Cunningham, in a paper read before the London Zoological Society in 1839, states that the bird was met with in all the wooded portions of the island. It reposed during

Brown Kiwi.   (*Buller.*)

the day, he says, in humid forests, living either beneath the tufts of long, sedgy grass, or shunning the light, and hiding itself in the hollows at the base of a rata tree. But no sooner were the native woods darkened by the presence of night, than it ranged about in quest of food. He describes the cry of the kiwi at night as being similar to the whistling of a boy by the help of his fingers in his mouth.

It was by imitating the cry that the Maoris decoyed the birds to their destruction. On the darkest nights the kiwi-hunters set forth, accompanied by dogs, and supplied with torches, so as to dazzle the birds' eyes. The kiwis go about in pairs, and the hunters endeavoured to secure the female first, easily distinguishing it by its large size. As the male lingered about the spot to protect its mate, both fell into the trap. Formerly the Maoris were skilful kiwi hunters. They delighted in the sport, "and many a group would they form to pass a dark, tempestuous night in the forest, in order to decoy and catch these birds."

The method of procedure in the hunt is related in the following words by a Maori friend of Mr. S. Percy Smith:—"The kiwi prefers a worm above all other foods. First comes one and then another kiwi, in search of worms, their heads always on one side, with an ear turned to the ground, listening for the creeping of the worm beneath the soil. As soon as one of them hears the creeping in the soil, down goes his beak, right to the worm, which it brings up to eat. The creeping noise of the worm in the soil is like that made by the hand of a watch, but rather louder. That is what the kiwi listens for. On account of this habit, the hunter carefully prepares little pieces of wood, which are tied to the dogs' necks, so that they may rattle as the animals move. Hearing this, the kiwi thinks that it is a worm, and stops to listen. While it is doing this the dogs are able to approach, and, by the time it starts to run, the dogs are baiting it. The men then advance with their torches, which are burning, and consequently the birds cannot see, and are caught and killed. The kiwis are never hunted by daylight." The skins of the birds were much sought after as material for mats, and the flesh was eaten.

The Southern Kiwi.—Rowi.

Apteryx australis.

Greyish brown, streaked longitudinally with black. All the feathers soft to the touch. Bill slightly curved. Claw on the wing slightly curved, and of a light horn colour. Eye, black. Male, length of the bill, 3.5 in. to 5.5 in.; of the tarsus, 2.25 in. to 3.5 in. In the female—Bill, 5.5 in. to 7.55 in.; of the tarsus, 3 in. to 3.5 in. Egg—White; length, 5 in. West Coast of the South Island and Stewart Island.

Apteryx australis, which was the first species made known to science, is called rowi by the Maoris, and big kiwi by the miners.

When irritated, it makes a cracking noise by snapping its mandibles together very rapidly. In attempting to defend itself, it displays an awkward feebleness rather than a posture of self-protection, by striking forward with its foot, as in the act of scratching, at a line about its own height, and its only defence against dogs is in concealment. In walking, it has a peculiar step. The foot is lifted deliberately, and rather high above the ground, and its gait is like that of a person who moves stealthily.

Between the sexes there seems to exist a lasting companionship, and it is thought that the birds pair for life. For a nesting place the rowi selects a hole in some huge tree or log, or amongst roots. Sometimes the hole is excavated in a soft bank, where the soil is light, but care is generally taken that the site shall be on a ridge of dry ground. The breeding season extends over some months, from October to February. Two eggs are generally laid, and the old birds rather lie than sit on them. The method of roosting is very peculiar. The birds squat opposite each other with their legs bent under them, each with its head tucked under the scanty apology for a wing. If there are young in the hole, they assume the same position on each side, there being a young one between the two parents, and the result of this singular family arrangement is a nearly perfect hemisphere of feathers. They often appear torpid or very drowsy when surprised in their homes, sometimes remaining quite undisturbed by noise. Their cry, which is harsh, and is something like "cr-r-r-ruck, cr-r-r-ruck," is not uttered until after sundown. The young are well clothed when they leave the shell. The general colour of the bird is greyish-brown, streaked with black in the young and adult state, and in some fine old birds a glint of golden chestnut edges part of the plumage.

The Grey Kiwi.—KIWI.

Apteryx oweni.

Grey, spotted with yellowish white, the spots forming bands. Bill, straight. Tarsus, pale brown. Claw on the wing, weak, slightly curved, and of a light horn colour. Eye, black. Male, length of the bill, 2.85 inches; of the tarsus, 1.75 inch. In the female, length of the bill, 3.5

inches; of the tarsus, 2.5 inches. Total length, about 19 inches. Egg—
White; length, 4.35 inches. South Island.

Apteryx oweni, which is sometimes called the straight-billed
kiwi, and the little grey kiwi by diggers, has similar habits to
those of the species just described. Mr. Potts says that he can see
no reason for mistaking the elaborately organised bill for an
instrument to be used like a pick for digging into hard soil. The

Grey Kiwi.
(*Rowley's Ornith. Miscel.*)

tongue is very short, but muscular, angular in shape, and it can
be used in crushing insects against the flat, opposed surface of
the upper mandible, as the strong muscle on the lower surface
gives a great degree of strength. The visual organs, which are
feebly developed, are placed so as to command the movements of
the upper mandible. The ears are well developed, and, as an aid
in discovering food, are next in importance to the olfactories.

The long straggling hairs, or weak bristles, planted among the
feathers of the anterior part of the head, fulfil the useful office

of protecting the eyes and head from injury, and may also guide or regulate the force of the thrust given by the bill. They form a perfect guard of feelers and a simple means of defence, in strict harmony with the retiring cautiousness which is the natural instinct of the kiwi. The under surfaces of the feet are well protected by cushions. The claws, which are slightly curved, are sharp at their points, and are admirably adapted for scratching, but are not shaped like those of the domestic fowl, which are adapted for traversing hard ground as well as for scratching. The robust tarsi, defended by hard scales, are articulated with the tibiæ by very strong joints, which must give the kiwi great power of leaping and jumping, and enable it to scale fallen trees, and search along their upper surfaces for insects. The hind toes and claws help in maintaining the position of the bird when fossicking about the prostrate trunks, strengthening the hold, and preventing the bird from slipping to the ground when reaching down.

The cry of this kiwi is described as like ''kvee, kvee, kvee,'' repeated sometimes twenty times in succession, with moderate haste. This scarcely ceases when it is replied to by ''kurr, kurr, kurr.'' These calls are heard through the night, sometimes commencing after sundown and ceasing about 3 o'clock in the morning.

The breeding season extends over several months, and eggs have been obtained on the West Coast during a great part of the year. The home of the bird is generally found beneath the spreading roots of trees, in logs, or under rocks, and contains one or two eggs or young, but no more. The nests are found on the bare soil, and are never constructed of dried ferns and grasses. The pair of birds usually remain together during some months, and share in the labours of incubation; but the male apparently allows much of the labour of rearing the young to be done by the female. The young have been found at a short distance from the family abode, in a kind of nursery. They are described as ''quaint looking little animals, with not too much of the savour of youth about them, being nearly exact miniatures of the adult. There is no young state of plumage with them, and none of that

half pronounced variation in tone, or tint of discolouration, which
calls for the nice discrimination of the practised ornithologist
when questions of age have to be settled. In winter and summer
alike, they adhere to their sober colours with quaker-like
pertinacity.''

It is thought that the separate lodging is set up until the young
are well able to forage for themselves, under the guidance and
protection of the old birds, so that the family party is not
necessarily broken up because all its members do not abide
together in one place of hiding and rest. There does not seem
to be any reason for believing that kiwis are great travellers,
as ample supplies of food may be obtained by fossicking around
their homes. Judging by the tracks made, they resort to the
same holes for some time, probably till the family has consumed
the more favourite kinds of food in the vicinity. These kiwis
seem to adopt the same squatting posture as the rowi, and are
quite as lethargic, allowing themselves to be captured without
further resistance than a feeble struggle, in which, at worst, a
scratch or two would punish incautious handling.

This species has been found very high up on the ranges, not
far below the snow; but, it is stated, always in the bush. Mr.
Potts took one from a deep hole beneath a fragment of rock, just
within the scrub bush, about a mile westward of the Franz Joseph
Glacier.

Dogs readily follow the scent of Apteryx. Very large numbers
of the birds have been needlessly destroyed by hunters. Bushmen,
it is stated, do not dislike the flesh of the kiwi, and the meat,
though coarse, has a gamey flavour. The eggs afford excellent
food.

This bird is especially abundant in the wild and mountainous
south-western portions of the South Island, and numbers of them
are kept at the Government sanctuary at Resolution Island. At
one time a large demand sprang up for kiwi skins, which were
used for making muffs, and the hunting of the birds became
almost a vocation. One kiwi hunter told Mr. Potts that up to the
close of 1871 he had killed about two thousand two hundred
specimens of *Apteryx oweni* and *A. australis*.

The Spotted Kiwi.

Apteryx occidentalis.

Similar to *A. oweni*, but rather larger, and the black bars on the feathers more distinct. Total length, about 23 inches. West coast of the South Island. This is a doubtful species.

(*Rowley's Ornith. Miscel.*)

Great Spotted Kiwi.

The Great Spotted Kiwi.—Roa-Roa.

Apteryx haasti.

Brownish grey, spotted with yellowish white, forming broad light bands. Bill slightly curved. Tarsi, dark brown; the claws dark. Eye, black. Male, length of the bill 4.75 inches; of the tarsus, 2.75 in. In the female, length of the bill 5.4 inches. Total length, 25 to 27 inches. Subalpine regions of the West Coast of the South Island.

This is a large bird. It was named as a compliment to the late Sir Julius von Haast.

The existence of an albino kiwi has been recorded. A specimen of this rare and beautiful bird was found in the bush near Martin's Bay, on the west coast of Otago. The plumage was white, but the extremities of the feathers were more or less stained with a yellowish tint. The bristly integument at the base of the mandibles was also yellowish, and there was a narrow yellowish stain round the eye, while the irides were brown and the feathers soft to the touch. Other specimens have been obtained near Greymouth.

INTERESTING PROBLEMS CONNECTED WITH NEW ZEALAND'S AVIFAUNA.

The avifauna of New Zealand, although very limited in numbers, contains so many peculiar species that it has always been an object of much interest to ornithologists.

Of the non-migratory land birds there are only one hundred and four species, belonging to sixty genera, and to twenty-four different families. Of these, eighty-three species, 80 per cent., and twenty-eight genera, 47 per cent., are found nowhere else; while there are no fewer than five families—Turnagridæ, Xenecidæ, Nestoridæ, Stringopidæ, and Apterygidæ—which exclusively belong to New Zealand.

This shows that there was long isolation from the rest of the world, during which time the species that originally came to New Zealand have been much modified, while in some cases their relatives, which were left behind in the old habitations, are now extinct.

Where were those ancient habitations?

By Dr. P. L. Sclater and Dr. A. R. Wallace, New Zealand is considered to be a sub-division of the Australian Region. But nearly one-half of the genera of land birds show no connection with Australia; and the only birds that show a distinctly Australian facies are the warblers, the tits, and the creepers, the three genera of honey-suckers (Meliphagidæ), and perhaps, the parrakeets. There is therefore, much to be said in favour of the opinion held by Professor Huxley and Professor A. Newton, that New Zealand should be kept as a distinct region.

There is much evidence to prove that New Zealand was at some former time, probably in the early Eocene Period, connected by a land-ridge with New Caledonia and New Guinea; and there is ample evidence to show that the Tasman Sea, which separates New Zealand from Australia, has been in existence during the whole of the Tertiary Era.

It is highly probable that the ancestors of many of the New Zealand birds came along this land-ridge from New Guinea.

The saddle-back and the huia, for instance, are starlings, allied to calornis, of India, the Malay Archipelago, New Guinea, and northern Australia.   The New Zealand thrush is much modified, and its nearest ally is Myiophoneus, of India and Java. The fern-bird is found in Southern Africa, with a near ally in Madagascar.   The New Zealand wrens are related to the pittas of India, the Malay Archipelago, New Guinea, and Northern Australia.   The wood pigeon has its nearest relatives in the fruit-pigeons of India, the Malay Archipelago, Polynesia, North Australia, and New Caledonia.   The kaka is a remarkable form, connecting the macaws, of South America, with the true parrots of Africa; but, as it formerly occurred in Norfolk Island, we may assume that it came to New Zealand from the north.   The wood-hen, also, has a close relation in Lord Howe Island.   And the blue duck has some connections in the Moluccas.

There is further evidence in the fact, that, at the present day our migratory birds still come to us from the north.

On the other hand, we know that the white-eye crossed over the Tasman Sea in 1856.   It appeared first in the south of New Zealand and worked its way north.   It is highly probable that the redbill, the shoveller, and the ancestors of the paradise duck did the same at some earlier period, for they are not found in New Guinea or in New Caledonia.

But, in many cases, it is difficult to decide from which direction the first birds came.

There are two possible explanations of the cause of the connection between New Zealand birds and Australian birds.

The ancestors of the present species may have crossed the Tasman Sea, and so have come to New Zealand from the west; or the relationship may be due to two branches of emigrants from New Guinea going southwards, one of which passed into Australia, and the other into New Zealand. In this category we have the warblers, the tits, the wood-robins, the creepers, the fantails, the honey-suckers, the ground-lark, the kingfisher, the parrakeets, the morepork, the quail-hawk, as well as several rails and ducks. Some of these may have crossed the Tasman Sea, but it is more probable that the ancestors of most of them came to New Zealand from New Caledonia.

There is good reason for thinking that a migration of land-birds into New Zealand from the north took place in the Eocene Period, for it is highly improbable that New Zealand was ever again connected with the mainland. In the two following Periods New Zealand stood at a lower level than at present by some 3,000 feet, but in the older Pliocene it had a much greater extent, and included the Chatham Islands, which were again separated from New Zealand in the newer Pliocene. If we knew in detail the history of the origin of the land-birds of New Zealand, it would enable us to solve several interesting problems in the development of species, or, at any rate, would throw considerable light upon them; but as that is impossible, we must do the best we can with the imperfect knowledge we have.

In the first place, we learn something about the relative ages of certain groups. For all those groups which are fairly well represented in the fauna, such as the parrots, the rails, the herons, the plovers, the ducks, and the kiwis, are, in all probability, old groups, unless they have crossed the Tasman Sea. The perching birds are also well represented, and, as they are mostly forest birds and poor fliers, we may suppose that most of them came to New Zealand in the Eocene Period. But the pigeons, the game-birds, and the Picariæ are each represented by a single species only; and it is, therefore, presumable that they have a later origin than the former. Among the perching birds, the absence of orioles, shrikes, flower-peckers, swallows, weaver-finches, and others, indicates that they are younger groups than the starlings, the honey-suckers, the fly-catchers, and the pittas. This argument, however, must be used with caution, because there may be other reasons for the absence of a group of birds from New Zealand than that it did not exist in the Eocene period.

Another subject of interest is the relative rate of change in different species. When we regard things on a large scale, we find some regularity in this connection. New Zealand, not including the outlying islands, has 90 species, of which 58, or 68 per cent. are endemic. The Chatham Islands have twenty-five non-migratory species, of which 10, or 40 per cent., are endemic; while the Auckland Islands have 14 species, of which 7, or 50 per cent., are endemic. Therefore, New Zealand, which has been separated longer from the mainland than the outlying islands have been separated from New Zealand, has a larger percentage of endemic species. It may be stated, also, that neither the Chatham Islands nor the Auckland Islands have any peculiar families; the Auckland Islands have, in the flightless duck, a peculiar genus, and the Chatham Islands had formerly four peculiar genera of rails.

But when we go into details, we come across many anomalies. The tits have varied between the two islands, but the South Island species is unaltered in the Chatham Islands and the Auckland Islands. Sir Walter Buller explains this by supposing that the birds in the Chatham Islands and the Auckland Islands are late comers, having flown over from New Zealand; but this explanation seems to be improbable, for if the bird was capable of flying to the Chatham Islands, it could certainly fly to the North Island, which it has not done. The fern-bird has changed in the South Island, where two species live together, but the tawny fern-bird is the only species on the Snares, while the Chatham Islands fern-bird is confined to those islands. The bell-bird has changed in the Chatham Islands, but not in the Auckland Islands. On the other hand, the tui has not changed on any of the islands. The North Island wood-hen remains the only species in the north, but in the South Island it has split up into three or four species.

As might have been expected, no good fliers have representative species in the two islands of New Zealand, but the bell-bird, the ground-lark, the parrakeets, the wood pigeon, and the sand plover are represented by different species in some of the out-lying islands. Some badly-flying species, such as the fern-bird, the warblers, the fantails, and the wrens, are identical on both islands of New Zealand. Some of the rails remain unchanged, while others have varied greatly. While the fern-birds, the tits, the warblers, and the fantails have undergone only specific changes since their ancestors came to New Zealand, the New Zealand members of the creepers (Paridæ) and of the honey-suckers have changed considerably. Enough has been said to show that not only do different species vary at different rates, but that the same species varies at different rates in different places.

New Zealand is well placed for studying the effects of environment on variation, but the study lands the investigator in puzzles. Certainly there is greater variety of conditions in the South Island than in the North Island, and the South Island has a larger number of species, but it seems to be impossible to connect the difference of conditions with any of the specific changes. Why should the crow have an orange wattle in the South Island? Why should the wood-hens be darker in the neighbourhood of the West Coast Sounds, and lighter in the sub-alpine ranges? Why should the fern-bird get paler in the south? What has the fawn-coloured breast of the rock-wren to do with living among rocks?

A special example may be taken. The green parrakeets, although not confined to New Zealand, have their headquarters here. There are three species living together in the South Island, two on the Antipodes, two on the Auckland Islands, and one each on the Chatham Islands, and Macquarie Island. The Auckland Island birds have not varied in plumage from the original stock in New Zealand, although they must have been isolated for a long time, as the crest of the breast-bone has undergone a reduction in size. The Antipodes Island parrakeet has varied considerably, but the Chatham Island and the yellowish parrakeet slightly; and it is impossible to connect these changes with the surrounding physical conditions.

If we cannot claim these variations as due to the action of the environment, neither can we claim them for natural selection.

z

We can understand how natural selection may have strengthened the bill in the thrush and elongated it in the huia, as we can see that both are useful; but we cannot see that the differences just mentioned are useful to the birds. Most of these cases are evidently connected in some way with isolation; but there are exceptions, as in the fern-bird, the fantails, the wood-hens, and the kiwis.

The case of the wood-hens is especially interesting. Here we find the North Island wood-hen in Stewart Island, but not in the South Island. In the South Island there are four other species, each of which is fairly well confined to its own district. The South Island wood-hen inhabits the low lands all round the islands; the hill wood-hen is found on the hills, the black wood-hen in the West Coast Sounds, and Finsch's wood-hen in the neighbourhood of Lake Te Anau. The South Island wood-hen is also found in Stewart Island, where it seems to interbreed with the North Island wood-hen, if we may judge from the number of individuals that come from there which are intermediate in plumage between the two. Although hybrids are very rare in a natural state, it is possible that Finsch's wood-hen may be a hybrid between the black wood-hens and the South Island wood-hens.

The only explanation of these facts seems to be the supposition that at one time the North Island wood-hen spread over all three islands, and that it remained unaltered in the North Island, while in the South it gave rise to the South Island and black wood-hens and then disappeared, the hill wood-hen being afterwards derived from the South Island wood-hen. But it is hard to say why these differences should have arisen.

Another very interesting case is that of the yellowish parrakeet, which has been found at the Antipodes Island and Macquarie Island. It differs from the red-fronted parrakeet by its yellow tint. It is extremely improbable that the parrakeet of Macquarie Island came from Antipodes, or *vice versa*. The islands are further apart than either of them is from New Zealand; and both Auckland and Campbell Islands, on which the species is not found, lie close to the track. The yellow tint is due to a failure of

the feathers to develop properly, and as the red-fronted parrakeet
sometimes shows this variation in New Zealand, it seems probable
that the yellow parrakeet originated independently in both islands
from different flocks from New Zealand. It is probable that the
small number of individuals which arrived at each of the islands
allowed the variation to obtain a permanent influence, which it
could not do in New Zealand, on account of the large number of
birds, and, consequently, the greater facilities for inter-crossing.
Thus, we have a new species produced by the isolation of a few
individuals. We also have a case of the double origin of the same
species. This last cannot occur often; but it shows that the great
differences between the climates and vegetation of Antipodes
Island and Macquarie Island failed to produce any effect.

New Zealand ornithologists have special advantages for studying the effects of the absence of enemies on development, the
most important of which is degeneration in the powers of flight.
No part of the world offers so many examples of degeneration in
the wings of birds as New Zealand does. There are strong flying
birds, such as the quail hawk, the kea, the parrakeets, and ducks;
as well as a chain of more or less degenerate birds, passing
through the tui, the thrush, the crow, the huia, and the fern-bird,
to the kakapo, the wood-hens, the flightless duck, and the kiwi,
none of which can fly at all. In these non-flying birds there are
some with large wings, like the kakapo, others with small wings,
such as the wood-hens and the kiwi, and there are no wings at all
in the extinct moas, in most of which even the shoulder-girdle had
disappeared.

In some of the outlying islands, the snipes, the larks, and the
parrakeets are feeble fliers, although their relations in New
Zealand fly well. So general an effect upon birds of so many
different kinds must be due to some general cause; and we find
it in the baneful effects of the want of competition. The fact that
on some of the outlying islands there are several birds with feebler
powers of flight than their congeners in New Zealand, although
they have no more enemies in New Zealand than on the islands,
is very remarkable. We may accept natural selection as the cause
of loss of powers of flight when that loss has been useful to the

species; as in the case of parrakeets and the larks on small islands much exposed to gales of wind, as the Antipodes. But this will not apply to the flightless duck of the Auckland Islands, or to any of the birds on the main islands of New Zealand. Who can doubt but that the kakapo and the wood-hens would be benefited by being able to fly?

The pectoral muscles of the tui have remained unaltered in the Auckland Islands, while in the parrakeets and larks they have been reduced. But the Auckland Islands are large, and have plenty of shelter, and it is not easy to believe that parrakeets and larks would be blown away from them. It is quite impossible to believe that the flightless duck should be in any such danger; so that, as the explanation of natural selection here fails us, we may doubt its effects in other cases.

As already stated, in New Zealand we cannot trace the action of natural selection in bringing about the degeneration of the wings of so many birds, some of which inhabit the forests and others the open country. It is in the absence of competition, combined with an abundant supply of food, and consequent disuse of the wings, that we must look for the cause of the loss of their powers of flight.

An examination of the birds shows that in most cases the crest of the breast-bone diminishes first, then the wings are reduced in size (first the feathers then the bones), until at last they disappear altogether. That is, the muscles get weak before they get smaller, while the wings remain large long after the muscles are too small to employ them efficiently.

How to account for this has always been a difficulty.

It has been suggested that natural selection, acting through the law of economic nutrition in periods of scarcity of food, would cause the degeneration of an organ which was of little or no use. The useless organ, it is thought, would not require nourishment, and would dwindle away, to the advantage of the rest of the animal. But for this explanation to hold good, it has to be established that periods of semi-starvation actually occur, and that when they occur, the useless organs suffer more than other parts of the body.

As to the periods of semi-starvation, it has to be stated that, among the perching birds, there are in New Zealand four genera of flycatchers—the tits, the wood-robins, the warblers, and the fantails—as well as the wrens, all of which live largely on insects; and it seems almost certain that these birds must be pushed very hard for a living every winter. There are also nine genera of seed-eaters, or grub-eaters; and these can have no periodical famine time, for the food is abundant all through the year. But none of the insect-eating birds show deteriorated power of flight, while among the others we have the huia, which seldom attempts to fly, the thrush, which cannot rise into the air without making several preliminary hops, as well as the crow, the fern-bird, and the tui, all of which are feeble fliers.

There is no reason for supposing that the wood-hens, the kakapo, and the kiwi have any periodical terms of semi-starvation, for all of them are as fat in the winter as in the summer. The parrakeets of the Antipodes Island and the Auckland Islands are as large and as well fed as those of New Zealand, although they have degenerated in the powers of flight. Lastly, as the moas were vegetable feeders, their food in a climate like that of New Zealand must have been independent of the season of the year. Consequently the hypothesis of the necessity for semi-starvation breaks down. Indeed, if a bird was perishing from periods of starvation, wings would be useful to enable it to search more ground, and they would be preserved by natural selection.

The evidence from New Zealand, therefore, is to the effect that degeneration is not primarily due to natural selection, either in the ordinary way or through a supposed law of compensation of growth.

The only explanation left is that degeneration is due to disuse-inheritance.

Disuse we know to be a true cause of degeneration in the individual—it is the only true cause that we know—and everyone will admit that it is the starting point, or originating cause, of degeneration. The difficulty is to explain how variations in an organ, caused by disuse, become progressive. That is, how do they accumulate, generation after generation, until at last the whole organ disappears? In other words how does disuse give rise to disuse-inheritance? This still remains a puzzle.

VOCAL CHARACTERISTICS OF SOME NEW ZEALAND BIRDS.

By T. H. POTTS.

(From *Out in the Open*.)

To those interested in the wilds of nature, much of the real history of bird-life is disclosed by their notes; for instance, if the voice of the kingfisher was heard from the first day of August to the month of January (the breeding season), it would not be necessary to see the bird in order to form a tolerably correct idea of the nature of its employment.

Bird-sounds, as received by the ear, it is impossible to reduce to writing, nor do I believe it will be achieved till science shall have instructed us by some method to render in intelligible language the many fleeting forms and figures which the Babel of tongues of sound impress on the wavelets of the surrounding air. Formidable discovery! Then we shall hold as a priceless truth that, if speech is silver, silence is golden!

But although it seems impossible to write down bird-sounds, yet a notion of their effect on the air-waves might be hazarded. For the purpose of explanation, let us suppose the existence of an undisturbed mass of air; could not the figures described therein by the calls of various birds be idealised into forms, and a symbolic rendering of the sounds of bird-language be produced?

As illustrating the meaning in view, let us suppose that the sharp jarring scream of the falcon would be represented by a figure somewhat like a barbed lance; the call of the cuckoo (*Chalcococcyx*) would be pictured in gentle sweeping curves; whilst an acute angle would typify the scream of the weka (*Ocydromus*).

THOMAS HENRY POTTS.
Born 23rd November, 1824.   Died 27th July, 1888.

From the notes and observations I have made, I have no doubt
that birds breed here in every month of the year; and according
to generally accepted opinion, therefore, we ought not at any
time to lose the music of the woods. But there are active agencies
at work which are quickly rendering whole districts comparatively
mute, and these will be presently touched upon.

At night we hear the sounds of birds high up in the air, as flock after flock seeks the coast or the brackish water of the shallow mere. These notes are probably, as Gilbert White said, a safeguard against dispersion in the dark, or may convey some intimation of any change in the order of flight; they are usually briefly, yet deliberately, sounded. Sea-fowl are far from silent when on their course, ascending rivers or roaming above the harbours and bays that indent the shore.

Living close to the beach in a sheltered nook in Port Cooper, at no great distance from the extensive area of Lake Ellesmere, it may be that I have been more than usually attentive to these wandering voices, since few woodland birds now frequent the slopes of our picturesque hills, like many other districts once clothed with stately trees and bright-leaved scrubs. Shade and shelter gone, bare stems with whitened tops remain, and point to the work of the ruthless bushman.

Often at night, about the second week in January, the shrill piping of the oyster-catcher (*Haematopus*) is heard, and, soon after, the yelping cry of the stilt (*Himantopus*), apparently from a great height. These waders are amongst the earliest to quit their inland breeding haunts and bring their pied broods towards the coast. They are on their way to join or assist in forming the large flocks which during the autumn and winter spread themselves along the shores and over flats and harbours, where abundance of food can be procured.

Many genera, which must in all fairness be termed gregarious, utter their calls and cries with frequent repetitions, and that too in broad daylight. Can we divine their meaning? Let us observe which are the noisy species. Flocks of tern may be heard screaming at some distance, as in open order and at no great height they stream across the country, foraging by sight. Is their squealing cry uttered in rivalry, for companionship, for encouragement, or satisfaction at the prospect of a well filled gullet? Watch a flock of the same species hovering over a river, and should anything unusual, such as a dead bird, be born down by the current, a clamour at once arises. How swiftly is the news spread from bird to bird! In a brief space hundreds are wheeling and screaming over the object of attraction. In this case the call conveys intelligence; it is analogous to the bushman's "coo-ey," attracting instant attention, and summoning the presence of all within reach of its sound. In the instances given, the call notes used appear very similar. By way of contrast, stroll across one of their breeding grounds when the down-clad young lie in couples without the slightest shelter. Fiercely is the intruder assailed; the harsh scream becomes intensified, and plainly expresses anger, defiance, and would-be intimidation, for the brave little tern protects its nestlings even against man, with a courage unknown to the most powerful gull. Our large gull (*Larus dominicanus*) will drive away the egg-stealing harrier, which soars aloft in wide circles on silent wing, as the gull chases it from the neighbourhood of the sandy shore or rocky cliff where the roughly-built nest protects the brown-blotched eggs; it marks each dashing stroke with a short bark of anger, and returns from the pursuit with hoarse, gratulatory noise. But when man assails its treasures, the miserable bird wheels aloft, and circling round in company with its neighbours, breaks forth into loud despairing cries that sound like thick-voiced mocking laughter. There is no *levée en masse*, as with the plucky terns; there is no attempt made to defy or inspire fear; but, securing itself from danger by ascending in wide circles, the loud-voiced sea-fowl looks down on the plunderer in timid helplessness, uttering incessantly its wailing lamentations.

Look at that flock of gulls which surround the shipping lying at anchor near the breakwater! What a busy picture of noisy activity! It is life at high pressure, and stands out in bold relief to the rest of the scene, where all around lies still and silent, steeped in the full glare of noon. Some are ranging restlessly in circles, and swiftly their shadows come and go upon the glancing waters; others sit lightly and gracefully on the rising swell—all on the look out for scraps that may be thrown overboard or swept through the scuppers of the ships. Suddenly one quick eyed bird pauses in its flight, hovers an instant, from beneath the snowy tail feathers drawing his pink feet, which for a brief space dangle in ungainly fashion ere they clutch the water; now he has snatched some bulky morsel. What a vociferous outcry, as half choked he strives to gulp it down! His wings, not yet close folded, he spreads again for flight. Attacked on all sides by his clamorous fellows, he drops the envied lump, and instantly joins the common flock in chase of the lucky bully that has swept off the prize. The pursued now becomes the pursuer, and the chase continues until some widely-distended throat at length entombs the object of this fierce contention. Here the birds among themselves, without man's interference, show an amount of boldness that appears remarkable; the air resounds with their sonorous cries. Seldom, if ever, is the hunted bird struck by his companions; he yields his prey from fear, or drops it in the attempt to obtain a fresh hold, and by another catch place it more easily for swallowing. If lost from fear, can it be from dread of the menacing blow that seldom, if ever, descends? Has the bird not instinct enough to appreciate the threatened attack at its true value, judging from its own harmless bullying?

On the mud flats at the head of the harbour, patched here and there with a dwarf growth of zostera, and banks of time-bleached shells, as the tide ebbs, flocks of godwits (*Limosa novae-zealandiae*) arrive, and probe the yielding surface with their long bills; their call cannot be distinguished from that of their European congener, although now and then a yelping sound is emitted without any apparent cause, unless it is a note of satisfaction, for they feed silently. Noisier, and far shriller in their notes are the oyster-catchers, which feed in company, wade in the shallow water, or course along the margin with swift splashing run. When the pied stilts feed in numbers by the shores of Lake Ellesmere, their notes are constantly repeated, sounding not unlike the barking of young dogs, whilst the oyster-catcher's shrill note rather resembles the running down of an alarum, in the rapidity with which the sound is repeated.

The call of the paradise duck (*Casarca*) is often heard in lofty flight, bringing to mind the notes of the wild geese at Home. Some fancy they can detect in the hoarse call of the paradise drake the words ''Hook it, hook it,'' as a hint to escape, whilst the shriller cry of the duck inquiringly replies, ''Where, where?''

Amongst other species which use the voice in company, and seem to enjoy the chorus, the lark may be mentioned, as it usually utters its sharp ''chirrup, chirrup,'' on taking wing. The same note, or one much like it, is used for encouragement, or to incite watchfulness when a flock in loose order are near a harrier hawking close to the ground, or perched on some commanding stone or tea-tree. When the blight birds (*Zosterops*), which might safely adopt as a motto "*Fruges consumere nati,*" crowd about a tree peering through the leaves, thrusting their sharp beaks into the flesh pulp of luscious plums, they constantly twitter, as they also do when shifting to fresh food; the call note, not unlike the chirrup of the sparrow, is also quickly answered. Their power of song, as yet, does not seem to be appreciated as it deserves. I have heard individuals sing their sweet few notes in a way that would charm the most exacting bird fancier that ever gathered chickweed. The notes of the bell-bird as it trips up and down the scale with a cough at the end are too well known to need further notice; one of their concerts, with a full chorus, is a delightful treat that sometimes rewards the early riser. The graceful parrakeet utters a gratulatory note as the flock hastily assembles to some favourite food, as on the stooks in an oat field; this differs from their call when on the wing, as much as it does from that low confidential murmur in which I have heard a pair indulge about nesting time. The kaka in his leafy domain utters his harsh grunt of satisfaction as he and his mate scramble about the bending boughs that yield a honeyed food. How shall I attempt to describe the song of the tui, with its sudden bursts of melody, ringing the change upon notes merry, plaintive, or harsh, in rapid sequence, as though the sympathetic voice felt and expressed every varying motion that chanced to stir the lively bird? The attitudes assumed during the course of its recitative are well worth watching, although they may seem to detract somewhat from the pleasure of hearing it.

The kaka sounds his alarm harshly, hopping restlessly from bough to bough; nor does his warning cease whilst on the wing, gliding to safer quarters. In the moist beech forests, where glades are carpeted with the deepest moss, the beautiful green wren sounds his cheepy cry, denoting danger, with a most confident air. Away out on the open ground or sandy riverbed, how often does the "twit, twit," of the banded dotterel, or sharply uttered "ti-winkle, ti-winkle" of the redbill or oyster-catcher, help to moderate the weight of the sportsman's bag; the paradise duck lifts its head, sounds his "kowonke," from a fast walk he hastens to a run, and at length sails away with his shriller-voiced mate.

Very noticeable is the faculty which birds possess of hushing their young to silence, and of bidding them hide at a moment's warning, perhaps by the sound of a single note. Amongst some species of waders this obedience to parental guidance is most observable; young stilts, plovers, or redbills, which have been rambling over their feeding ground, at the sound of alarm suddenly seek cover, and only after the most careful scrutiny may be found lying *perdu* behind some sheltering stone. Perhaps the most monotonous amongst all the calls of our young birds is that of the large gull (*L. dominicanus*). When nearly fully grown, about the months of April and May, it follows the old bird with untiring perseverance, clamouring for food with a long squealing cry. I have heard it on the beach whilst it has been wheeling round and round to reach its parent's bill in hopes of a supply, till the sound has become quite tiresome to listen to. By way of contrast to the patience of the old gull, it may be noted that the young of the tit, when well grown, as it is by December, is driven off by both parents with something like harshness both in tone and gesture. The fierceness which is displayed by the common tern (*S. albistriata*) in defence of its young has already been noticed; a similar degree of courage is to be met with in the case of the falcons and the little grey warbler (*Pseudogerygone igata*). On nearing a taratah (*Pittosporum eugenioides*), where some yellow warblers were perched, the old birds commenced a furious attack, darting close to the face, precisely after the manner of the common tern, and, allowing for size and power, uttering a similar jarring scream to that bold bird. With the falcons the utmost perseverance is exhibited in driving away a foe. In December, up the gorge of the Lawrence, a pair of bush hawks (*Nesierax australis*) assailed one of my sons and myself for a space of two hours, whilst in the neighbourhood of their young; then the usually swiftly uttered "kli, kli, kli, kli" was even more rapidly sounded, whilst its tones were savage and threatening. The young at the time were able to fly some little distance, yet only one moved once, that we could observe, from the instant the note of alarm was given. The bronze-winged cuckoo or whistler (*Chalcococcyx*) always makes known his presence with an oft repeated whistle; the long-tailed koekoea announces his arrival with deep-breathed note; these love calls are unlike all others of our bird sounds. The wild scream of the weka tells us of his whereabouts from a considerable distance; and this most confident of rails is as noisy by night as it is by day. When sitting still in the forest, I have seen a weka silently approach, and give notice of my presence by a strange note, which, although delivered within a few feet of where I was sitting, sounded like wood being struck at a great distance off.

The remarkable notes of the owls must not be passed over silently; for the name at least, if not the appearance, of the morepork (*Ninox novae-zealandiae*) is well known throughout the colony. Australian settlers distinguish a *Podargus* by a similar name, whence the colonial epithet (whether of New Zealand or Australian origin is uncertain) applied to a dawdling person, who is often described as "a regular old morepork." The call of the whekau (*S. albifacies*) is vociferous, wild, often startling from their heavy slumbers the inmates of the mountain huts. Probably the clamour of this genus, like that of *Falco*, is a means of startling some of their prey into motion. The large owl is said to have likewise a call somewhat similar to the morepork, but much more gruff in tone. Laughing jackass is one of the names conferred on the whekau; this distinction is shared by an Australian bird as well as by some of our sea birds amongst the petrels or *Procellaridae*.

When the south-east wind blows on our east coast, bringing with it thick hazy weather, when curling mists drift up the harbours and hide away in their vaporous mantles hill and mountain, shearing the landscape of its fair proportions, the curious note of the petrel may be heard from dusky eve till early morn, not only about the harbours and estuaries, but far up the riverbeds to the gorges in the vast mountain chain of the Southern Alps.

Amongst the most silent of our birds may be named the shags (*Phalacrocoracidae*), the harrier, the heron, and the grebe, whose voices, except during the breeding season, are rarely heard. The squeal of the harrier is not infrequent, considering what a very common bird it is. In the breeding season the scream is heard from a bird soaring high in air, or frightened from its nest, or suddenly driven off its prey, occasionally only from a bird on the wing hawking the burnt ground, which has disclosed, perhaps, an unusual abundance of lizards. The cries of birds in several cases appear to be more or less dependent upon atmospheric changes. At such times gulls become vociferous, restless, soaring aloft with rapid unsteady course, and wekas are very noisy: on the other hand, many species are silenced altogether by bad weather. The thrush, of many notes, utters some so like those of other birds as to become rather puzzling should one try to fix on the unseen performer. The flute-like mellow pipe of the wattle bird (*Glaucopis*) is unrivalled for its sweetness. The little creeper (*Acanthidositta*) never moves without emitting its tiny twitter. The kingfisher is generally silent, except during the breeding season, or its note is used to intimidate, either when attempting to seize a post already occupied by one of its kind, or when defending its position from an attempted intrusion; thus our kingfisher differs in habit from that of the Old Country, which is said to utter its cry whenever it takes wing. Notwithstanding the gush of song which in summer-tide salutes the cool dawn, before the rosy hues have fired the eastern sky, many of our little melodists retire late to rest, such as *Anthornis, Petroeca, Pseudogerygone,* and *Zosterops,* and their lingering notes may be heard long after sundown. Often is observation made upon the readiness with which some species of our native birds learn to imitate the human voice, an accomplishment which is always popular; yet, as an exhibition, the result of long practice and frequent repetitions, I am inclined to place it in the same category as a man's imitation of the crowing of a cock. Some persevering enthusiasts find that the kaka, parrakeet, and tui are the most apt to acquire this power of uttering sounds that bear a fancied resemblance to words.

In the foregoing notes, the voice of the large gull has been more than once mentioned. On the mudflats or sandbanks, when a small flock of five or six of these birds are met together, after a few deep toned barks or growls, they hold a regular "tangi," and utter most dismal wails or yells, or what seems like a dialogue or discussion takes place, very often received by the auditory with mild barks that might well pass for applause, or "loud and continued cheers." This habit is not confined to the large gull, but is also possessed by the smaller species, *tara-punga*, although the latter is less noisy. The terns, too, meet in parliament on the shore; and a solemn conclave of oyster-catchers may sometimes be noticed standing in unusual repose, at intervals only uttering a shrill pipe, and this when, close at hand, the godwits are working in their tripod fashion to extract a dainty morsel from the ooze.

Attention has already been directed to the fact that, in the alpine districts of New Zealand, the notes of the birds are pitched in a higher and richer tone than in the valley, and in some of the most elevated woods which the bell-bird frequents we have found the note or brief song of the hen bird specially delightful. Whence this result? To the quality of its food being climatically altered? If we notice some of the fruits and berries from which it derives some portion of its support, we shall find that the blackberries of *Aristotelia racemosa* are represented in the alpine fastnesses by those of *A. fruticosa*, the pulpy fruit of *Coriaria ruscifolia* by that of *C. thymifolia* and *C. angustissima*, whilst the drupes of *Coprosma lucida* and those of many other species have their mountain representatives in *C. cuneata, C. acerosa, C. linariifolia,* and others. Will the chemist tell us from analysing these fruits that this change is enough to cause some modification in the muscular apparatus that modulates the tones issuing from the syrinx? The scientific ornithologist would admit no specific difference after inspecting a score of skins; for length of feathers, colour of plumage, point out the bird as *Melanura*.

As to the reason for the bell-bird's song being pitched in a higher key, it may perhaps be found in the fact that thick mists often envelop the mountain's side, that the bushes in the more elevated gullies are much scattered, small, and isolated. Hence the alpine note is fitted to meet the peculiar physical conditions of certain localities, by enabling the sexes to communicate with each other when collecting food at some distance apart.

The power of imparting intelligence, as exercised by birds, must be obvious to anyone who is acquainted with the ordinary inmates of a poultry-yard. In many feral species that have come under observation, this faculty is quite as conspicuous as it is amongst many domesticated *protegés*. One summer, for the first time, a few tuis appeared amongst the cherry trees in a garden up the gorge of the Ashburton, miles away from any wood frequented by the tui; for the first time cherries were tasted, and the knowledge of their excellence was communicated, and the trees stripped by the industrious tuis. Not a very great time ago, when slowly sailing up the harbour, one of the children threw a piece of bread to a young gull (*L. dominicanus*), the only bird in sight. Its bark of pleasure brought others, till then unseen, and the wake of our boat was enlivened by an irregular train of noisy attendants. Those species which do not launch lightly in the air when taking flight, I believe, may be ranked among the more silent birds, as, for instance, the cormorants; birds of the genus seem to need a fulcrum to rise upon the wing. The fleet kingfisher, too; when its perch is a bough, and it leaves it to dash at its prey, the bough may be seen to vibrate for some time after it has been quitted. Both of these genera may be fairly classed with the non-vociferous tribes, notwithstanding that the Halcyon indulges in a variety of expressive notes during the breeding season.

I now leave with regret the interesting study of bird sounds, and trust that others will prosecute further observations, for there is much to be learnt by the field naturalist about their notes and calls, which would assist in revealing many interesting points in the history of the fauna.

REPTILIA

Cold-blooded vertebrates, which breathe air all through life.

ORDER LACERTILIA.

Head, and usually the body also, covered with scales. Eyelids always present. Legs usually four. Teeth fixed to the jaw-bone.

Family Geckonidae.

Scales on the head, the rest of the body soft, with small granules on the back, and small scales on the abdomen.

Key to the Genera.

1. Last joint of the toes rising from within the digital expansion.	Gehyra.
Last joint of the toes at the end of the digital expansion.	2
2. Toes gradually narrowing.	Naultinus.
Toes abruptly narrowing.	Dactylocnemis.

Genus Gehyra.

Toes strongly dilated, the end-joint free, raised from within the extremity of the dilatation. Inner toes without claws. India, Australia, Polynesia, and Mexico.

The Oceanic Gecko.

Gehyra oceanica.

Above brown, either uniform or with lighter and darker markings; below uniform whitish. Length, about 6 in. Moluccas, New Guinea, Polynesia. Specimens have been obtained on Flat Island, Moko Hinou Group, north of the Great Barrier Island.

(*Voy. Coquille.*)

Oceanic Gecko.

Genus Naultinus.

Toes feebly dilated, gradually narrowing, with a series of transverse lamellæ under their entire length. Colour green. New Zealand only.

The green geckos live on shrubs and in open fern land. They hybernate in the winter, from June to September, six or seven being huddled together. They are viviparous, bringing forth from one to three, usually two, young ones at a time, between July and September. In the summer they cast their skin several times.

They feed upon insects, and are very abstemious. Mr. Colenso, who kept several alive for more than a year, found that feeding them with flies twice a week was quite sufficient. He says: "Their manner of taking their prey is peculiar. When the lizard clearly sees the fly, and makes sure it is living, it steals towards it in the most stealthy manner. As it nears the fly, and when within two inches, then is the time to watch closely its actions. First it arches its neck to a tolerably sharp angle, and its eyes swell and bulge out, or rather upwards, over their orbits, and the expression of its countenance alters greatly, taking on a fierce look. It lifts its little hand-like paws, and moves them, only a toe or finger at a time, and often in the air, very slowly and cautiously, much as a little child moves its hands when stealing along on tip-toe. Then it nears its head towards its prey, but so very slowly that I have better detected its movement by watching its shadow cast on marked paper by strong sunlight, reminding me of the almost imperceptible movement of the hour-hand of a clock. At last it has got to about one inch, or a little less, from the fly. As quick as light the dart is made, and the fly is caught; and the little lizard rapidly knocks about its prey from side to side, as a terrier with a rat, not, however, striking the fly against anything, but merely shaking it. After a short time spent in that manner, the lizard proceeds to swallow the fly, which it does by half opening its mouth and drawing it up, and generally, after three or four movements of this kind, the fly is gulped down whole, legs and wings and bristles."

"It is pretty to see them drinking," he adds; "this they do but seldom; they lap water much like a cat, but very slowly, as if they were tasting it, every now and then passing their broad, thin, and large tongue right over their eyes, as if washing them, and always so finishing the drinking. Their tongue and palate are of a purple colour. They seem to like the water, as they often go singly into their water-trough, and remain extended in the water for some time. They swim very fast, but clumsily, as if they were in a great hurry about it."

They are not timid animals, and allow themselves to be caught easily. They can run swiftly when they try, with an undulating movement of the tail. The tail is prehensile, and is used in climbing trees. It is stated that they assume all manner of curious and grotesque positions. Whatever posture they assume they can keep for a long time, and they remain motionless for hours, occasionally even days, in one position.

Sir W Buller says that this lizard, on being molested, emits a peculiar chattering sound, which the Maoris term laughing, and of which they have a wide-spread superstitious dread. The laugh of the green lizard was enough to terrify the bravest warrior, and its occult power for evil was strangely believed in by all the

tribes in every part of the country.   The reptile itself, whether
dead or alive, was an object of universal fear among them.
Mr. Colenso says that he never heard its cry.

Key to the Species.

1. Skin of the back rough.   N. rudis.
   Skin of the back smooth.   2
2. Back with pale spots or lines. N. elegans.
   Colour of back uniform.   N. grayi.

The Spotted Lizard.—KAKARIKI.

Naultinus elegans.

Upper surface green, with two series of yellow or white spots, which
are sometimes margined with black, on the back and tail. Lower surface
yellowish white. Length, about 4½ in., of which more than half belongs
to the tail. Very variable in its markings. Sometimes the spots are
replaced by streaks, and rarely there is a yellow streak down the middle
of the back. Usually there are some spots on the head. Both islands of
New Zealand.

Although this lizard is most commonly found on the low land,
a specimen has been obtained by Mr. Brough under stones among
the snow near the top of Mount Arthur, in the district of Nelson.
It was only 2.7 in. long.

The Green Lizard.—KAKARIKI.

Naultinus grayi.

Upper surface green, sometimes with minute black dots. Lower surface
yellowish white. Length, about six and a half inches, of which the tail
occupies more than one half. North Island only. Common.

The time of gestation of this lizard must be long, at least five
and a half months, as a specimen kept by Mr. Colenso, which was
caught on December 29th, 1885, gave birth to two young ones
on June 8th, 1886. These young ones were dead, and each was
wrapped in a semi-transparent membrane, showing that the birth
was premature. They measured about three-quarters of an inch
in length, and were green on the back, shaded off in spots to
lighter green, and almost to white in some of the little knobs and
slight hollows. The mother was entirely green, and was six
inches and a half in length. A yellow variety has been found

at Maketu, Taranaki, and the Kaipara. The young of this variety are green on the back.

Green Lizard.

(Voy. Beagle.)

The Rough Lizard.

Naultinus rudis.

Head and back covered with small granular scales, intermixed on the sides with large, roundish, flat or keeled, tubercles. Greenish grey above, with irregular longitudinal and transverse purplish bands on the back. Uniform light grey beneath. Length, 5¾ inches, of which the tail occupies more than half. Very rare.

Genus Dactylocnemis.

Toes dilated at the base, suddenly narrowed, the narrow portion forming an angle with the basal portion; series of transverse lamellæ under the latter. Colour brown. India and New Zealand.

Very little is known about the domestic economy of the brown geckos. They are nocturnal in habits, and hide under stones or the bark of trees in the daytime.

Key to the Species.

1. Compressed portion of the toe ⅓ of the whole.	D. maculatus.
Compressed portion of the toe ½ of the whole.	2
2. Toes webbed at the base.	D. pacificus.
Toes not webbed at the base.	D. granulatus.

The Tree Lizard.—MOKOPAPA.

Dactylocnemis pacificus.

Brown above, with irregular transverse bands on the back and tail, and frequently a dark band on each side. Lower surface, whitish. Toes not much dilated, the width of the dilated part being one-third of its length; the length of the narrow part is one-third of the whole; a distinct web on the base of the toes. Length, 6½ inches, of which the tail forms about one-half. North Island only.

The tree lizard lives on trees, and is very sluggish in its movements. It is common in the Auckland district.

The Long-toed Lizard.

Dactylocnemis granulatus.

Greyish, or brown, above, with dark brown vermiculations and irregular cross-bands, which are light edged in front; a dark band on each side. Toes not much dilated, and no trace of a web; the length of the narrow part is one-third of the whole. Length, up to 7½ inches, of which the tail forms more than half. Both Islands, also Stephen Island.

Of this species Mr. Colenso says:—''I obtained two fine living specimens of this lizard last summer while in the woods, and one since, a smaller one, also living, from Mr Balfour. This last is still living, although it has not eaten anything since I received it, nearly six weeks back. It has only taken, at intervals of several days, a very little water, and this when I put it into a wash-hand basin to take a swim. On being taken out, it invariably licks up a few drops. Hitherto it has refused flies, which my other lizards always greedily ate; and I have supposed much might be owing to its hybernating season not being over. It is exceedingly quiet, and rarely moves about. It often changes its ground-colour of grey to a pink-red, and this it does sometimes three or four times a day; the cause, however, of its doing so is entirely unknown to me. I have often tried, by altering its position as to light, and to sun heat, and also by giving it a little gentle shaking, to see if I could cause it to change colour, but I have never once succeeded. It seems to be entirely dependent on itself, and not arising from any outside cause, nor from the time of day, neither is it regular in its changes.''

Long-toed Lizard.

(*Brit. Mus. Cat.*)

The Short-toed Lizard.

Dactylocnemis maculatus.

Brown above, with small blackish spots and more or less distinct, irregular transverse dark-brown bands on the back and tail; a dark streak passing through the eye; sometimes a broad light band on each side of the back. Under surface dirty white. Toes considerably dilated; length of the narrow part, one-fourth of the whole. Length, about 6 inches, of which the tail forms rather more than half. South Island and Stephen Island.

This is the common brown gecko of the South Island, where it is found generally in the bush, but not uncommonly under stones, especially where bush has formerly existed. Nothing has been recorded about its habits.

Family Scincidae.

Head, body, and tail covered with scales. Eyelids well developed, movable, scaly, or with a transparent disc.

Short-toed Lizard

(*Brit. Mus. Cat.*)

Genus Lygosoma.

The characters of this genus are entirely osteological, and cannot be given here. It will be sufficient to say that all the New Zealand skinks, or true lizards, belong to it. Asia, Africa, and North America.

These lizards are diurnal in habit, and move swiftly, so that they are difficult to catch; but no one appears to have observed them carefully.

Key to the Species.

1. Soles of the feet black.   L. grande.
   Soles of the feet yellowish.   2
2. Tail forming much more than half the length.   3
   Tail forming about half the length.   5
3. Less than 32 rows of scales round the body.   L. moco.
   More than 32 rows of scales round the body.   4

The Rock Lizard.

Lygosoma grande.

Black above, spotted with pale olive, or olive dotted with black. Lower surface, greenish or pale olive. Scales of the feet black. Tail, about once and a half the length of head and body. Forty to fifty rows of scales round the middle of the body. Length, 8 or 9 inches. Found among rocks, chiefly in the southern part of the South Island. Sometimes the lower surface is reddish orange.

Rock Lizard.   (*Brit. Mus. Cat.*)

The Common Lizard.—Mokomoko.

Lygosoma moco.

Brown or olive above, with a more or less regular, black-edged, light. dorso-lateral streak. Back uniform, or spotted with blackish and yellowish, or with dark brown, black-edged longitudinal lines. Lower surface, yellowish, greyish, or pale olive; uniform or spotted with black. Soles of the feet, yellowish. Tail, about one and a half times as long as head and body. There are 28 to 30 rows of scales round the middle of the body. Length, about 7 inches. Common throughout both islands of New Zealand.

The mokomoko is found under stones and logs, both in the bush and in the open country. The colours are very variable. Generally the back is pale, or dark reddish brown, with a whitish stripe, edged below with black, from the nostrils over the eye down the side; and often with a dark stripe down the centre of the back. Sometimes it is pale golden brown on the back, and darker on the sides, with a narrow black line dividing the colours. A dwarfed form occurs among the shingle-slips of the mountains of the South Island.

Common   Lizard.   *(Gray's Liz. of Aust.)*

Pitt Island Lizard.

Lygosoma dendyi.

Dark olive grey above, with small black spots and a blackish brown wavy lateral band passing through the eye; this band sometimes dotted with white. Under surface, leaden grey or blackish. Tail, once and a half as long as the head and body. There are 34 rows of scales round the middle of the body. Length, about 6½ inches. Pitt Island, Chatham Group.

The Long-tailed Lizard.

Lygosoma lineo-ocellatum.

Back, pale brown or greyish olive, with small black spots or ocelli with a white centre. A more or less marked whitish dorso-lateral band. Sides variegated with blackish, or blackish dotted with white. Lower surface, greyish; the throat and breast black spotted. Tail, one and three-quarter times as long as the head and body. There are 32 to 36 rows of scales round the middle of the body. Length, 8 or 9 inches. South Island of New Zealand and Stephen Island.

Ornamented Lizard. *(Gray's Liz. of Aust.)*

The Short-tailed Lizard.

Lygosoma smithii.

Brown or olive brown above, usually with small black spots, and a more or less distinct light dorso-lateral band edged below with brown. Sides paler, often with light dots. Lower surface, yellowish or pale olive; the throat usually variegated with blackish. Sometimes the whole animal is black. The tail is about as long as the head and body. There are 36 to 38 rows of scales round the middle of the body. Length, about 4½ inches. Both islands of New Zealand, generally on shingle beaches.

The Copper Lizard.

Lygosoma aeneum.

Brown above, with a few darker dots; a dark brown irregular dorso-lateral streak, edged above with yellowish. Sides with light, dark-edged spots. Throat spotted with dark brown. Tail thick, about as long as the head and body. There are 26 to 28 rows of scales round the middle of the body. Length about 5 inches. Both islands and Stephen Island.

The Ornamented Lizard.

Lygosoma ornatum.

Yellowish or reddish brown above; each scale with several fine darker lines. Sides, brown, with dark brown and yellowish markings. A yellow, dark-edged spot below the eye. Lower surface, yellowish; uniform or spotted with brown. Tail, thick, about as long as head and body. A small transparent disc is sometimes, but not generally, present on the lower eyelid. There are 28 to 30 rows of scales round the middle of the body. Length, about 5 inches. Auckland district. Very like the last species.

ORDER RHYNCHOCEPHALIA.

Lizard-like animals, but with considerable differences in the skeleton, among which are a well-developed breast-bone and abdominal ribs. They are considered to be the most generalised of all living reptiles, approaching the amphibia. Their affinities are equally close to the turtles and to the lizards; and they form one of the oldest known reptilian types. In the newer Cretaceous Period an aquatic form lived in North America and Europe.

Genus Sphenodon.

Scales on the upper surface small and granular, intermixed with small tubercles; those of the lower surface large and transverse. No ear openings. Tail compressed. Toes webbed at the base. A low crest along the neck, back, and tail. Two parallel rows of teeth in the upper part of the mouth. New Zealand only. The nearest ally to Sphenodon is Homœosaurus, which lived in Europe in the Jurassic Period.

The Tuatara.

Sphenodon punctatus.

Yellowish, or greenish olive with yellow spots. Length, up to 20 in. The female is larger than the male, and generally darker in colour.

Formerly, the tuatara lived in large numbers on both the main islands of New Zealand, but it is now seldom found anywhere

except on a few islets lying off the coast: The Chickens and the
Little Barrier, in Hauraki Gulf; Karewa Island and the Rurima
rocks, in the Bay of Plenty; Motiti and East Cape Island; and
The Brothers and Stephen Island, in Cook Strait.

Dr. Dieffenbach, writing in 1843, says that he had been
apprised of the existence of a large lizard, which the natives
called tuatara, or Ngarara, as a general name, and of which they
were much afraid. But although he looked for it at the places

Tuatara.   (*Voy. Erebus and Terror.*)

where it was said to be found, and offered great rewards for a
specimen, it was only a few days before his departure that he
obtained one, which had been caught at a small rocky islet called
Karewa, about two miles from the coast in the Bay of Plenty.

The tuataras live in holes in the ground, sometimes in company
with a petrel of some kind, and can be got out only by digging.
Their food is principally beetles, grasshoppers, and spiders, but
they will eat any small animal so long as it is alive. Captain
Mair states that he caught some tuataras on the Rurima rocks.
At the same time he put a number of small lizards into a box with
them. There were at first about twenty of them. He observed
that they diminished in number every day, till at last only six

remained, and these appeared to be quite paralysed with fear. Whenever the tuatara made a movement, the poor little creatures would crouch down and try to hide themselves under the dry leaves in the box. He watched the box very carefully, and at length found a tuatara in the act of eating one of the small lizards. It had crushed the little fellow quite flat, beginning at the head, and rejecting about an inch of the tail. At the bottom of the box he found a dozen tails. There were four or five little tuataras, about three inches long, in the box, but none of those had disappeared. He generally found them perched on the heads of the big ones, asleep.

Tuataras are very capricious in their feeding, sometimes refusing it for months, and then suddenly eating heartily every day. They are fond of water, and like to lie in it for a large part of the day; and they swim freely, sometimes with only the nostrils above water, sometimes immersed altogether.

Dr. Newman has described these animals as being lazy in all their movements, and, when frightened, he says, they move very slowly. He adds that their pace is a slow crawl, the abdomen and tail trailing on the ground. When driven fast, or when chasing prey, they always lift the whole trunk off the ground; it does not touch at any point. This rapid gait is very wobbling, something like a man swimming sideways. After running three or four yards, they grow weary and stop. They cannot jump the smallest obstacle, their limbs being too feeble.

They sleep the greater part of the day, and awaken slowly even when handled. They sometimes bite when caught, and occasionally fight viciously with each other, and then they bite hard, holding on tightly with their mouths. They are timid animals, easily frightened by noises, and run into their holes at the sight of a man. They make no sound.

They dig holes for themselves in loose sand or shingle, using their fore feet, sometimes right and left alternately, sometimes one foot only. The earth is thrown far behind them. They grow slowly, and live for a long time. Captain Mair says that he has seen an unusually large one which had been kept in an old kumara pit on Flat Island, Motiti, for over three generations.

Dr. Dendy's inquiries show that their eggs are laid in holes, which they specially dig for the purpose. The eggs are white and soft, with a semi-calcareous shell, and are about one inch long. From eight to ten are laid at a time. They are laid in November, and do not hatch until about midsummer in the following year. Development goes on in the egg during the summer, but in March it almost ceases, and is resumed in the spring. The earlier stages of development resemble those of the tortoise, and it is only later that its more lizard-like characters appear. On the snout there is a sharp-pointed shell-cutter; and the body is marked with longitudinal and transverse bands of grey and white, a pattern completely lost in the adult animal, which is spotted.

In captivity, they will eat live worms or thin shreds of raw beef, if these foods are hung up in the cage, and are very fond of live snails. They live for a long time without any food. A paragraph published in the *Lyttelton Times*, in 1903, stated that "the tuatara lizards at the Opawa fisheries seem to be susceptible to music. They will come out of their holes in the rocks to hear a song, when nothing else will induce them to appear. They prefer a good rousing chorus rather than a solo. Some time ago, a number of visitors to the hatcheries wanted to see the tuataras, which, however, refused to come forth, until a little girl sang 'Soldiers of the Queen,' and others joined in the chorus. The sound seems to have appealed to the reptiles, and they responded by showing themselves to the singers "

It is stated that tuataras are by no means as stupid as they seem to be. In some instances they display a rather surprising amount of intelligence. They recognise people readily, and are sensitive to the presence of strangers. They have to be handled carefully, as they can give a nasty bite when in a bad temper. In Auckland, a gentleman had his thumb bitten by a female tuatara he was lifting up to show to a friend. He caught the reptile by the body, and with a quick movement of her head she nipped a piece off his thumb.

AMPHIBIANS

Cold-blooded vertebrates, which breathe water by means of gills in the early part of their life, and then change to air-breathers. The skeleton is on the same type as that of the reptiles, but there are no ribs, or only very small ones. Fins, if present, have no fin-rays.

ORDER ECAUDATA.

In the perfect state, four limbs and no tail.

Family Discoglossidae.

Upper jaw with teeth; the vertebræ with short ribs.

Genus Liopelma.

Tongue, circular, entire, free behind. Pupil triangular. No tympanic disc. Fingers free. Toes webbed; the tips not dilated. New Zealand only.

The New Zealand Frog.

Liopelma hochstetteri.

Upper surface with small smooth tubercles; lower surface smooth. Brown; limbs and lower surface lighter. Head and back indistinctly spotted with blackish. Limbs regularly cross-barred with blackish. Male without vocal sac.

The New Zealand frog has been found in the Coromandel Peninsula; at Huia, on the north side of the Manukau Harbour; and at Opotiki. In the early days of mining at the Thames, several were caught in the drives, where they had wandered during the night.

The habits and life history of this frog have been described by Mr. G. E. Archey, of the Canterbury Museum.* He states that at Coromandel it is found very rarely near streams; but occurs fairly abundantly high up on the crest of the Tokatea Ridge, where, in summer there are no pools or streams in which it might breed. It has however, attained independence of water

(*Vehr. zool.-bot. Ges. Wien.*)

New Zealand Frog.

1. Side view.   2. Lower surface.   3. Side view of open mouth.   4. Open mouth.   5. Fore-foot.   6. Hind-foot.

*Records of the Canterbury Museum, Vol. 2. No. 2, 1922.

by its eggs, which are laid on the ground in clusters of from six to twelve, being enclosed each in a transparent gelatinous capsule containing a clear fluid. The tadpoles, which have a large yolk-sac, go through their development within this small bath, breathing by means of a large vascular tail, and hatch out as soon as both pairs of legs are developed, in about a month's

(*From a drawing by Phyllis F. Clark.*)
Stephen Island Frog.

time. The yolk-sac and tail are not absorbed until the end of another month. Although surface water is absent from the frogs' habitat, sea mists continually drive over the crest, and so keep the area damp enough for the frogs' maintenance, and also probably have rendered possible the evolution of their specialised life history.

Liopelma belongs to a small family of frogs which inhabit Europe and northern Africa, Asia, and California, where a species occurs on glaciers. Liopelma probably came to New Zealand by a former land connection between this country and south-eastern China.

Stephen Island Frog.

Liopelma hamiltoni.

Slenderer than previous species; fingers and toes much longer, and webs much more reduced; tongue much narrower and not so free posteriorly. Form moderately robust. Skin smooth in almost all parts, but scattered tubercles on proximal portions of thighs and shanks. Colour—Light brown, with irregular darker and lighter marbling on the upper surfaces.*

This, the latest addition to the land vertebrate fauna of New Zealand, was discovered by Mr. H. Hamilton, a member of the staff of the New Zealand Dominion Museum, on Stephen Island, in Cook Strait, in 1918. The island had been occupied by lighthouse-keepers for many years before the discovery was made. In the same year Dr. J. Allan Thomson, Director of the Dominion Museum, visited the island, and found fourteen specimens of this frog crowded under a large heap of stones. Further specimens were found later, but the species, unless it is present on other islands, or on the mainland, is rare. Its life history is not known, but the climate of Stephen Island is similar to that of Coromandel, so it is not unlikely that this species has a modified development like that of the preceding species. Stephen Island is an islet, and is notable as one of the homes of the tuatara, which finds a refuge there.

---

* Description by Mr. A. R. McCulloch, Zoologist, Australian Museum.

APPENDICES

MAMMALIA

ORDER CARNIVORA.

Teeth sharp, covered with enamel; the canines long in both jaws.

Family Canidæ.

Limbs well developed; digitigrade; claws not retractile; muzzle produced; teeth, 38 to 46.

Genus Canis.

Teeth usually 42. Fore feet with five toes, hind feet with four toes; the claws slightly curved and blunt, widely spread.

Maori Dog.—KURI.
Canis familiaris, variety *maorium*.

About the size of a collie, but the legs shorter. Length of the skull, 145 to 179 mm. Width of the zygoma, 87 to 112 mm. For measurements see "Trans. N.Z. Inst.," vol. xxx., p. 152.

The Maori dog is extinct beyond all possibility of doubt. It was a dull, stupid, quiet, lazy, sullen, and very ugly dog, almost without a bark. When alive it was mostly a plaything and pampered pet to its Maori owners; when dead, its flesh was used as food, its skin for clothing, and its hair for ornaments. As it was almost the only representative of land mammals known to the Maoris before the arrival of Europeans, its importance was magnified. It was small, and had a pointed nose, pricked ears, and very small eyes. In colour it was black, brown, or particoloured, and had long hair and a short bushy tail. Some dogs were favourites of Maori women, who nursed them with affection. The species, apparently, was not indigenous to New Zealand, but was brought to the country by Maoris at the time of the migrations from the Pacific Islands.

Mongrels, crosses between the Maori dog and species introduced by Europeans, were mistaken for the true Maori dog.

They went into the forests and mountains and became wild, a condition into which the Maori dog never lapsed, although its stupidity often caused it to be lost. These mongrels were known as "wild dogs." They attacked settlers' sheep, and occasionally attacked men like packs of wolves. The wild dogs were exterminated some time before 1900.

ORDER RODENTIA.

Incisors large and chisel-like. No canines.

Family Muridæ.

Genus Mus.

Incisors narrow, without grooves. Ears and eyes large. Muzzle naked. Thumb with a short nail. Tail long, nearly naked.

Maori Rat.—KIORI.

Mus exulans.

Greyish white on the back, paler below. Tail as long as the head and body. Ears large and round. Length of head and body 4.75 to 5.5 inches; tail the same length.

The Maori rat was believed for many years to be indigenous to New Zealand, but Captain Hutton abandoned his name, *Mus maorium,* and it now bears the name of *Mus exulans,* of Polynesia. The Maoris evidently brought it to New Zealand in their canoes, with the Maori dog. There is nothing remarkable about this little rodent except strange swarming habits. Members of the species have made extraordinary migrations, swarming in thousands up hill and down dale, along valleys, and through villages, following, apparently, a mysterious Pied Piper. Their migrations were noted mostly in the northern part of the South Island. A notable one, in December, 1884, is described in *The Transactions of the New Zealand Institute,* volume xvii., by Mr. J. Meeson. On that occasion a large part of the extreme north of the South Island was subjected to an invasion and a plague. The Maoris relished the flesh of this rat, which was systematically snared by them. Their methods are described by Mr. Elsdon Best in *The Transactions of the New Zealand Institute,* volume xli. (1908). There is no reason to believe that the species is extinct, but it is some years now (1924) since a specimen was reported.

AVES.

INTRODUCED BIRDS.

Most introduced small birds were brought from the Old Country to deal with insect pests, which threatened early settlers with ruin. A few, notably the skylark and robin redbreast, were brought because they recalled pleasant memories of the country the settlers had left. The skylark succeeded beyond all expectation; robin redbreast utterly failed to establish itself Later, several species, notably the pheasants and the quails, were introduced as game birds. About 130 species of birds have been introduced, but only about 25 have become absolutely established in a wild state. The following list includes most introduced birds which have had an influence in the Dominion, sentimentally or otherwise, and which have become established.

Canadian Goose.

Branta canadensis.

Greyish-brown; paler below, or whitish-grey; head and neck black, with broad white patch on throat and around each side of the head; tail black, with white upper coverts; bill, legs, and feet black.

Efforts to establish these birds have been made by the Wellington Acclimatisation Society, and Southland and North Canterbury Societies, also by the Government. Some liberated at Glenmark, North Canterbury, in 1907-8 have established themselves. They have succeeded also at Lake Sumner, North Canterbury, Lake Te Anau, Southland, and Lake Hawea, Otago, and appear to be doing well in several other parts of the Dominion; they are reported to be increasing on Lake Ellesmere and Lake Coleridge. On Lake Ellesmere now, 1923, there is a flock of about five hundred.

Black Swan.

Chenopsis atrata.

Brownish black above, paler below; primary and secondary feathers white; lores pink; bill pink, banded with white.

Black swans are present in colonies of thousands on lakes, estuaries, and lagoons in many parts of the Dominion. The Avon River, Christchurch, Lake Ellesmere, and Lake Forsyth, Canterbury, and Lake Ianthe, South Westland, may be specially mentioned on the mainland. On some of the large lagoons on the Chatham Islands, 470 miles east of the mainland, black swans are amazingly plentiful. They nest amongst vegetation close to the margins of the water. The nests sometimes are made on floating platforms constructed of reeds in shallow water near the shores. The platforms are made by bending down some reeds, and placing others on them until a solid, almost circular, structure is formed, several inches above the surface of the water. The nest is a large shallow cup in the centre. Usually five eggs, greenish-white or greyish-green, nearly 4½ inches long, are laid. The black swan is classed as a game bird in the Animals Protection and Game Act, and an open season may be declared for shooting it under license. Except for that, it is protected. The North Canterbury Acclimatisation Society has permission from the Government to take black swans' eggs at Lake Ellesmere. The society lets contracts for collecting the eggs, usually to fishermen who live near the lake, and it sells them to confectioners. All eggs taken are stamped with the society's stamp, and it is illegal for anybody to be in possession of eggs without authority. The society sometimes deals with 1,500 dozen eggs from Lake Ellesmere in a season.

The following is an account of black swans' nests on Lake Ianthe:—

"Black swans nest on the banks or in shallow water close in. On land their nests are approached by low archways and beaten paths in raupo or flax, and are not more than a few yards from the water's edge. The cygnets reach the water down the slopes without any trouble whatever. Floating nests are on bulky platforms, disc-shaped or roughly oblong. Bulrush roots, used largely, seem to have been pulled from the bottom of the lake in shallows. A nest with five eggs, found on October 26th, apparently had no owner. When inspected two days later it still was unoccupied, the swan evidently having deserted it. Nests in the lake are sufficiently light to float. The platforms are three feet in diameter, sloping up from the edges to the centre, which is about eight inches high. There a slight depression holds the white eggs tinged with green, more than four inches long, usually five in number. The platforms are made of bulrushes, raupo leaves, and even twigs compactly tramped, but they are fixed by the use of live stems, which are anchors. This does not prevent the cygnets' cradles being rocked in every wavelet. Mr. J. Petersen, surfaceman, who has watched swans make floating nests, states that they begin by bending several live bulrushes together. Dead stems and other material are placed on the bent stems, and a platform is made. He has seen swans continuing the work after the eggs were laid, probably to reinforce the structure or repair damage by wind and weather.''*

White Swan.

Cygnus olor.

Pure white; bill reddish orange, with black knob at base; lores black; legs and feet black.

The white swan, unlike the black swan, has not succeeded. Twenty years ago it was one of the ornaments of the Avon River in Christchurch, but it never was plentiful there, and it has completely disappeared from all the reaches of the river in the city. There are a few in the lower reaches in Burwood and New Brighton, the birds, apparently, nesting on the shores of the Estuary. At present, early in 1923, means are being considered to induce white swans to return to the upper reaches.

Pheasants.

Phasianus colchicus and P. torquatus.

Phasianus colchicus.—Male: Crown of head bronze-green; mantle, chest, breast, and flanks fiery orange; feathers of upper back and scapulars mottled with black and buff; lower back, rump, and upper tail-coverts maroon, glossed with purple and green; tail feathers olive-green with narrow, wide-set black bars. Female: General colour sandy brown barred with black; tail feathers with wide irregular triple bars of black, buff, and black.

Phasianus torquatus.—Male: Crown of head bronze-green; general colour of lower back, rump, and upper tail-coverts greenish or bluish-slate colour, with rust-coloured patch on each side. White ring round neck. Female resembles female of P. colchicus; no white collar. In neither species has the female the red face-wattles of the male.

---

* ''Lyttelton Times,'' November 22nd, 1922.

Both the common pheasant (*P. colchicus*) and the ring-necked pheasant (*P. torquatus*) succeeded at first wherever they were introduced. One of the early successes was achieved by Sir Frederick Weld, a Premier of New Zealand, who established the common pheasant in Canterbury in 1865. Three years later, the North Canterbury Acclimatisation Society bred forty common pheasants, and sold them for £2 a pair. The extensive Cheviot Estate, between Christchurch and Kaikoura, soon was stocked with that species. The two species seemed to be spreading over the country, when they were unaccountably checked. They then began to decrease. They now have disappeared almost completely from the South Island, and are seen in only a few districts—in the Urewera Country, for instance—in the North Island. This result is attributed to the laying of poison for rabbits, depredations of stoats, weasels, and wild cats, bush fires, and to the pheasants' food supplies being taken by introduced small birds. It is stated that 150 crickets were taken from the crop of one pheasant. Gardeners and agriculturists in districts in which pheasants were plentiful complained bitterly of the mischief they did. They destroyed young grass, pulled up maize, and attacked potatoes, carrots, beans, wheat, and many species of fruit. A farmer at Parua Bay, Whangarei Harbour, in 1906, described pheasants as "the greatest curse settlers have to contend with."

Australian Quail, Swamp Quail, or Brown Quail.

Synæcus australis.

Male: Upper parts reddish-brown on sides, dull grey down middle, with a few fine mottlings of black; on under parts the buff feathers are grey down the centre. V-shaped black bars on feathers almost obsolete; sides of throat and head dull grey. Female: Black markings and patches on upper and under parts much coarser; shaft-stripes much wider than in male, and pale buff.

Fruit-growers complain of this quail in the same way as they complain of the pheasants. It is stated also, that quail take the seed and young clover in bush clearings. Never plentiful in the South Island, and now (1923) almost completely disappeared from it. Fairly plentiful in parts of the North Island. This quail often is mistaken for the native quail, which has not been reported for many years.

Californian Quail.

Callipela californica.

Male: Back, sides of neck, mantle, and chest grey; lower back, rump, and upper tail coverts olive-brown, washed with grey, or grey, washed with olive; chin, throat, and cheeks black; well developed crest of black club-shaped feathers. Female: Crest shorter and browner; no black and white pattern on head and throat. General plumage of both sexes with scaly appearance.

The Californian quail shares the enmity of farmers, on account of its taking seeds and fruit. Very plentiful in Taranaki Province and other parts of the North Island; also in the northern districts of the South Island, but not common in other parts of the South Island. A farmer states that it gets on very well in land where there is plenty of forest. Mr. D. Hope, the North Canterbury Acclimatisation Society's curator, states that it is fairly plentiful on Canterbury riverbeds where broom, gorse, and other shrubs supply sufficient cover, and that it feeds largely on the seeds of broom, gorse, and lupin.

Little Grey Owl.

Athene noctua.

Greyish-brown above, with white markings; white with brown streaks below. Smaller than the native morepork.

The little brown owl was introduced into Otago and Canterbury to check the small birds. It has established itself in both provinces, particularly in Otago, where good reports are issued of it. In 1909, it was reported from Alexandra, Otago, that the owls had given good service in coping with the small birds. The first consignments were liberated in Otago in 1906, and in Canterbury in 1908. The first birds for Otago were obtained from Germany, and the species is known in New Zealand sometimes as the little German owl.

Skylark.

Alauda arvensis.

Upper parts warm brown, mottled and streaked with darker; throat whitish or buff; under parts yellowish white.

The skylark is accused of pulling up springing wheat, and is troublesome in gardens when turnips, cabbages, and other early seeds are sown. Its song is heard in suburbs and open country districts, also above the extensive shingly beds of wide snow rivers in the South Island. It was introduced for sentimental reasons. From that point of view it is an unqualified success. When hard facts are dealt with, it is placed next to the sparrow in respect to destructiveness. Plentiful in all cultivated districts; fairly common in some suburbs. Nests in fields and dry river-beds.

Song-thrush.

Turdus musicus.

Olive brown above; breast yellowish, spotted with dark brown; throat ashy white.

The best songster in New Zealand. Takes fruit, insects, snails, slugs, and earth-worms. Nests in and frequents many suburban gardens. Destroys snails by hammering them on stones, as in the Old Country. Is seen occasionally in the interior of forests, where it nests in leafy shrubs. Mr. J. Joyce, Bligh's Road, Papanui, a suburb of Christchurch, wrote on January 3rd, 1923 :—

"I don't know if you are aware that the notes of the song-thrush differ with different members of the species. There are several song-thrushes in my garden, and I have made it a study to become familiar with the notes of the best songster and compare them with the notes of others. I find that the others do not use the notes I am accustomed to hear from my favourite. I often was curious to learn if all song-thrushes used the same notes; I am satisfied that each individual has his own notes. Song-thrushes begin to sing regularly every morning about a quarter past three. One begins, then another, and others follow, until some six or seven provide the melody. They continue until there is sufficient light for them to pick up their breakfast, generally of snails. They hammer the shells on bricks or stones, and get at the flesh. When

they have retired after their matins, blackbirds appear with their soft melodious notes. They are more timid than the song-thrushes.''
Mr. J. W. Roberts, a member of the staff of Messrs. Whitcombe and Tombs, Christchurch, supplies the following corroboration of Mr. Joyce's note:—''The Rev. Frank Gorman, better known as 'The Singing Parson,' while on a visit to Christchurch, was walking with a friend along the Riccarton Road, when he remarked on the beautiful song of a thrush in the trees near by. Returning to Christchurch twelve months later he took the same walk with his friend, and again the thrush was heard. Mr. Gorman remarked that that was the same bird they had heard on a previous occasion. His friend doubted the statement, saying that there were plenty of song-thrushes in the vicinity. Mr. Gorman said: 'No; my friend, no two birds have exactly the same note, and a trained musical ear would quickly detect a variation; that is the same bird.' ''

Blackbird.

Turdus merula.

Male: Plumage all black; bill and orbits of eyes orange-yellow. Female: Plumage sooty-brown.

Condemned by orchardists. It takes currants, strawberries, apricots, cherries, gooseberries, plums, peaches, pears, apples, and other fruits. It is accused of spreading the blackberry in some districts. Destroys insects and worms. Very plentiful in town and country, where its song is a delight.

Hedge-Sparrow.

Accentor modularis.

Back and wings reddish-brown streaked with dark brown; belly buff white; sides of neck, throat, and breast bluish or steel grey.

The hedge-sparrow is not a sparrow, but a member of the famous family that includes the nightingale and the red-breast. English ornithologists have tried to get the name of this bird changed, but have not succeeded, and hedge-sparrow, apparently, it will be as long as the English language lasts. It is a favourite of orchardists, as it takes many insects, particularly the green-fly aphis, and does no harm, or very little. Plentiful in almost all open country. Some county councils pay for the hedge-sparrow's eggs, under the impression that it is an injurious bird.

Sparrow (male in foreground)

Starling

Blackbird

Cirl-bunting

Yellow-hammer

Goldfinch

Rook

Skylark

Australian Magpie.

Gymnorhina leuconota.

Male: Generally black, but whole of the upper surface except outer parts of wings, and band on tail, white; head black; bill bluish-white to bluish-black at tip; legs black. Female: Same as male.

Described by farmers as a very useful bird. Probably the most useful bird introduced into New Zealand. Plentiful in certain districts in the North Island from Wellington to Whangarei, also in parts of North and South Canterbury. Reported now to be decreasing in Taranaki.

Rook.

Corvus frugilegus.

Black, with purple and violet reflections; base of bill, nostrils, and space near bill white.

Rooks have been established at Fendalton, a suburb of Christchurch, at Lake Taupo, at Hawke's Bay, at Hick's Bay, at North Canterbury, and other places. In 1918, Hawke's Bay farmers sounded an alarm by charging rooks with attacking sheep and lambs, and causing heavy losses, but the charges do not seem to have been authenticated. Complaints by agriculturists are more general, and are supported by clear evidence. In the New Zealand Parliament in August, 1923, rooks were vigorously condemned by experienced farmers. Some farmers, however, welcome rooks. They usually stay in one district, and do not spread readily, but sometimes change the sites of their rookeries. One of the oldest and largest rookeries in New Zealand is in tall bluegum trees in Fendalton Road, Fendalton, near Christchurch.

Starling.

Sturnus vulgaris.

Male: Black with purple and green reflections, upper feathers tipped with pale buff; bill yellow; feet flesh colour, tinged with brown. Female: Like the male.

Some farmers have described the starling as the only introduced bird in New Zealand worth having. It is found in almost all parts, and on the Campbell and other islands. In Canterbury farmers erect nesting-boxes to encourage starlings

Redpoll (male on right)

Hedge Sparrow

Greenfinch

Chaffinch

Song Thrush

Ring-necked Pheasant

Lapwing

Common Pheasant

to come about their farms. Starlings take insects and ticks off sheep and cattle, and destroy skylarks' eggs. As in other countries, they take fruit, but not in large quantities. Reports are received occasionally of extraordinary scenes at flights and gatherings of starlings. Mr. W. W. Smith, New Plymouth, reports that "tens of thousands of starlings perform their cloud-like gyrations around and above the island of Moturoa. It is truly magnificent to see them; they form densely black cloud-like masses."

House-myna.
Acridotheres tristis.

Head, nape, upper part of hind neck, chin, and throat black, feathers of head and nape long and narrow, forming a crest; hind neck, back, sides of chest, and flanks buff brown; bill yellow; tail brown-black, but tips of all but centre tail-feathers white; edges of wing-feathers white. Tail becomes remarkably abraded, the white tips sometimes completely disappearing.

The house-myna or minah, an Indian bird introduced to New Zealand from Australia, takes many insects, but also attacks fruit, particularly cherries; and in Hawke's Bay, where it is plentiful, orchardists complain of its mischief. It is plentiful in only a few other districts, notably Taranaki.

House Sparrow.
Passer domesticus.

Male: Throat and breast black; back chestnut-brown, streaked with black. Female: Without black on the throat, and upper parts streaked with dusky brown.

Although the introduction of the sparrow usually is condemned, sometimes immoderately, there is no doubt that it did good service to the New Zealand agriculturist and horticulturist in former days, when caterpillars and other insects were very troublesome, so much so that people feared that they would be eaten out of house and home. A new generation has arisen, and only the sparrow's faults are remembered now, except by a few. A correspondent at Tinui, Castle Point, east coast of the Wellington Province, wrote:—"Sparrows do a great deal of good. I have known them to clear a field of peas of caterpillars which, before the birds became plentiful, would have destroyed all the peas.'' Plentiful in almost all parts except dense native forests.

Chaffinch.

Fringilla cœlebs.

Male: Back reddish; rump green; breast chestnut; wings black, with two white bars on each; some wing feathers tipped with yellow. Female: Brownish, with conspicuous white wing-bars.

The chaffinch joins other finches in taking seeds and fruit, as well as caterpillars and other insects. It is not as popular in New Zealand as in the Old Country. Seen occasionally in city and suburban gardens. Very plentiful in many farming districts, and in some bush districts.

Redpoll.

Linota rufescens.

Male: Crown and breast crimson or carmine; red or pinkish-red on rump; flanks fawn colour; belly dull white. Female: Duller and no red on breast or rump.

The redpoll is not as plentiful as some other finches, but seems to be steadily increasing, reports often coming in of its appearance in districts. Early in 1923, Mr. W. Cobeldick, the Tourist Department's ranger, reported it from the Rotorua district. A very useful insectivorous bird, although reports from the North Island state that it is destructive on grass-seed burnings. Takes large quantities of the green-fly aphis. Often seen in back country, far from civilisation.

Goldfinch.

Carduelis elegans.

Forehead and throat blood red; cheek and part of neck white; back of head black; back brown; wings black and white and largely yellow.

The goldfinch takes large quantities of thistle-seed, and has not aroused as much enmity amongst farmers as some other small birds have. Plentiful in some districts, also present on Campbell Islands.

Greenfinch.

Ligurinus chloris.

Yellowish green; wing feathers with bright yellow outer margins; below greenish-yellow. Female duller.

Described as the farmer's enemy, particularly when grain is ripening. Destructive to fruit in some districts. Very plentiful in open country; Chatham Islands.

Cirl-bunting.

Emberiza cirlus.

Male: Top of head green streaked with dark; chin and throat black, succeeded by bright yellow gorget; cheeks yellow, with black line through eyes; breast olive-grey; belly yellow; flanks streaked with darker. Female: Much duller, no black and yellow facial markings; no yellow gorget.

The cirl-bunting is distinguished amongst the introduced small birds by its black and yellow costume. It is common at Hawera and other Taranaki districts, and at several districts adjacent to Christchurch, although not reported in the city. It seems to have created neither good impressions nor bad ones.

Yellow-hammer.

Emberiza citrinella.

Head, throat, breast, and under parts bright yellow, mostly spotted or streaked with brown; upper part brown, streaked with darker. Female: Duller, plumage spotted with reddish brown.

The yellow-hammer sometimes is mistaken for a native bird, the Yellow-head, described in the early part of this work. Flocks of yellow-hammers occasionally are seen in suburbs, and even in towns. The species is plentiful, and fairly troublesome in some country districts. In bush districts they often select stables as their haunts, nesting in shrubs and trees in the vicinity. Farmers and bushmen know the species as "the finch." They have headquarters at Mr. W. Wright's stables, close to the road at Pukekura, South Westland, in the heart of a great forest, also at Mr. W. Goss's stables, in heavy forest on the banks of the Arnold River, near Lake Brunner, Westland.

INDEX OF MAORI NAMES

INDEX OF COMMON NAMES

INDEX TO INTRODUCED BIRDS

SCIENTIFIC DIVISIONS

CLASSES.

ORDERS.

FAMILIES.

Genera and Species.

TABLE OF NEW ZEALAND AIR-BREATHING VERTEBRATES

(The numbers refer to the pages.)

F. Ziphiidæ: 55.
    Berardius arnuxi (Porpoise Whale).
    Ziphius cavirostris (Goose-beak Whale).
    Mesoplodon (Scamperdown Whale).

F. Delphinidæ: 57.
    Delphinapterus leucas (White Whale).
    Cephalorhyncus hectori (Porpoise).
    Delphinus delphis (Dolphin).
    Orca gladiator (Killer Whale).
    Globicephalus melas (Black-fish).
    Tursiops tursio (Cow-fish).
    Prodelphinus obscurus (Bottle-nose).
    Grampus griseus (Risso's Dolphin, "Pelorus Jack").

II. AVES: 64.
  SUB-CLASS CARINATAE:
  1. O. Passeres: 65.
    (a) Sub-order Oscines:
    F. Corvidæ: 65.
      Glaucopis wilsoni (North Island Crow)
        ,,   cinerea (South Island Crow).

    F. Turnagridæ: 69.
      Turnagra crassirostris (South Island Thrush).
        ,,   tanagra (North Island Thrush).

    F. Sylviidæ: 75.
      Pseudogerygone igata (Grey Warbler).
        ,,   albofrontata (Chatham Island Warbler)
      Petrœca macrocephala (Yellow-breasted Tit).
        ,,   toi-toi (White-breasted Tit).
      Miro albifrons (South Island Wood Robin).
        ,,   australis (North Island Wood Robin).
        ,,   traversi (Black Wood Robin).
        ,,   dannefordi (Little Wood Robin).

    F. Muscicapidæ: 85.
      Rhipidura flabellifera (Pied Fantail).
        ,,   fuliginosa (Black Fantail).

    F. Sturnidæ: 87.
      Heteralocha acutirostris (Huia).
      Creadion carunculatus (Saddle-back).

F. Timeliidæ: 95.
  Sphenœacus punctatus (Fern Bird).
      ,,   fulvus (Tawny Fern Bird).
      ,,   rufescens (Chatham Island Fern Bird).

F. Paridæ: 98.
  Mohua ochrocephala (Bush Canary).
  Certhiparus albicapillus (Whitehead).
  Finschia novae-zealandiae (Brown Creeper).

F. Motacillidæ: 104.
  Anthus novæ-zealandiæ (Ground Lark).
  Anthus steindachneri (Antipodes Lark).

F. Meliphagidæ: 106.
  Zosterops cærulescens (White-eye).
  Prosthemadera novæ-zealandiæ (Tui).
  Pogonornis cincta (Stitch-bird).
  Anthornis melanura (Bell-bird).
      ,,   melanocephala (Chatham Island Bell-bird).

(b) Sub-order Clamatores:
  F. Xenicidæ: 120.
    Xenicus longipes (Green Wren).
        ,,   gilviventris (Rock Wren).
    Traversia lyalli (Stephen's Island Wren).
    Acanthidositta chloris (Bush Wren).

2. O. Halcyones: 125.
  F. Alcedinidæ: 125
    Halcyon vagans (Kingfisher).

3. O. Cuculi: 129.
  F. Cuculidæ: 129.
    Chalcococcyx lucidus (Shining Cuckoo).
    Urodynamis taitensis (Long-tailed Cuckoo).

4. O. Psittaci: 138.
  F. Nestoridæ: 138.
    Nestor meridionalis (Kaka).
        ,.   notabilis (Kea).

F. Psittacidæ: 147.

Cyanorhampus unicolor (Antipodes Island Parrakeet).
,, cyanurus (Kermadec Island Parrakeet).
,, novæ-zealandiæ (Red-fronted Parrakeet).
,, erythrotis (Yellowish Parrakeet).
,, auriceps (Yellow-fronted Parrakeet).
,, forbesi (Chatham Island Parrakeet).
,, malherbei (Orange-fronted Parrakeet).

F. Stringopidæ: 151.
Stringops habroptilus (Kakapo).

5. O. Raptores: 157.
F. Falconidæ: 157.
Nesierax novæ-zealandiæ (Quail Hawk).
,, australis (Bush Hawk).
Circus gouldi (Harrier).

F. Strigidæ: 165.
Sceloglaux albifacies (Laughing Owl).
Ninox novæ-zealandiæ (Morepork).

6. O. Columbiformes: 171.
F. Treronidæ: 171.
Hemiphaga novæ-zealandiæ (Wood Pigeon).
,, chathamensis (Chatham Island Pigeon).

7. O. Galliformes: 175.
F. Phasianidæ: 175.
Coturnix novæ-zealandiæ (New Zealand Quail).

8. O. Ralliformes: 178.
F. Rallidæ: 178.
Hypotænidia philippensis (Pectoral Rail).
,, macquariensis (Macquarie Island Rail).
,, muelleri (Auckland Island Rail).
Cabalus modestus (Mangare Rail).
Nesolimnas dieffenbachii (Dieffenbach's Rail).
Ocydromus earli (North Island Wood Hen).
,, brachypterus (Black Wood Hen).
,, finschi (Finch's Wood Hen).
,, australis (South Island Wood Hen).
,, hectori (Hill Wood Hen).

F. Rallidæ—*continued.*
> Porzana affinis (Marsh Rail).
> ,,   tabuensis (Swamp Rail).
> Porphyrio melanonotus (Swamp Hen).
> ,,   chathamensis (Chatham Island Swamp Hen).
> Notornis hochstetteri (Takahe).

9. O. Heriodiones: 198.
    F. Ardeidæ: 198.
> Herodias timoriensis (White Heron).
> Notophoyx novæ-hollandiæ (White-fronted Heron).
> Demiegretta sacra (Blue Heron).
> Ardetta pusilla (Little Bittern).
> Botaurus pœciloptilus (Bittern).

10. O. Limicolæ: 205.
    F. Charadriidæ: 205.
> Arenaria interpres (Turnstone).
> Hæmatopus longirostris (Oyster Catcher).
> ,,   unicolor (Red Bill).
> Charadrius dominicus (Spotted Plover).
> Ochthodromus obscurus (Dotterel).
> ,,   bicinctus (Banded Dotterel).
> Thinornis novæ-zealandiæ (Sand Plover).
> ,,   rossi (Auckland Island Sand Plover).
> Anarhynchus frontalis (Wry-bill).
> Himantopus leucocephalus (White-headed Stilt).
> ,,   picatus (Pied Stilt).
> ,,   melas (Black Stilt).
> Recurvirostra novæ-hollandiæ (Avocet).

    F. Scolopacidæ: 221.
> Limosa novæ-zealandiæ (Godwit).
> Limonites ruficollis (Red-necked Sandpiper).
> Heteropygia acuminata (Sandpiper).
> Tringa canutus (Knot).
> Gallinago aucklandica (Auckland Island Snipe).
> ,,   huegeli (Snares Snipe).
> ,,   pusilla (Chatham Island Snipe).

11. O. Gaviæ: 228.
    F. Stercorariidæ: 228.
> Stercorarius crepidatus (Skua Gull).
> Megalestris antarctica (Sea Hawk).

F. Sternidæ: 231.

    Hydroprogne caspia (Caspian Tern).

    Sterna albistriata (Black-fronted Tern).

        ,,   vittata (Swallow-tailed Tern).

        ,,   frontalis (White-fronted Tern).

        ,,   fuliginosa (Sooty Tern).

        ,,   nereis (Little Tern).

    Procelsterna cinerea (Grey Noddy).

    Micranous leucocapillus (White-capped Noddy).

    Gygis candida (White Tern).

F. Laridæ: 237.

    Larus dominicanus (Black-backed Gull).

        ,,   scopulinus (Red-billed Gull).

        ,,   bulleri (Black-billed Gull).

12. O Tubinares: 242.

    F. Procellariidæ: 243.

        Oceanites oceanicus (Wilson's Storm Petrel).

        Garrodia nereis (Grey-backed Storm Petrel).

        Pelagodroma marina (White-faced Storm Petrel).

        Cymodroma melanogaster (Black-bellied Storm Petrel).

    F. Puffinidæ: 245.

        Puffinus bulleri (Long-tailed Shearwater).

            ,,   chlororhynchus (Wedge-tailed Shearwater).

            ,,   gavia (The Shearwater).

            ,,   obscurus (Dusky Shearwater).

            ,,   assimilis (Allied Shearwater).

            ,,   carneipes (Pink-footed Shearwater).

            ,,   tenuirostris (Tasmanian Mutton Bird).

            ,,   griseus (New Zealand Mutton Bird).

        Priofinus cinereus (Brown Petrel).

        Priocella glacialoides (Silver-grey Petrel).

        Majaqueus æquinoctialis (White-chinned Petrel).

            ,,   parkinsoni (Black Petrel).

        Œstrelata macroptera (Grey-faced Petrel).

            ,,   lessoni (White-headed Petrel).

            ,,   nigripennis (Black-winged Petrel).

            ,,   cervicalis (Black-capped Petrel).

            ,,   neglecta (Kermadec Island Mutton Bird).

            ,,   inexpectata (Rain Bird).

            ,,   cooki (Cook's Petrel).

            ,,   axillaris (Chatham Island Petrel).

        Ossifraga gigantea (Nelly).

F. Puffinidæ—*continued.*
    Daption capensis (Cape Pigeon).
    Helobæna cærulea (Blue Petrel).
    Prion vittatus (Whale Bird).
        ,,   banksi   ,,
        ,,   desolatus   ,,
        ,,   ariel   ,,

F. Pelecanoididæ: 267.
    Pelecanoides urinatrix (Diving Petrel).
        ,,   exsul   ,,

F. Diomedeidæ: 267.
    Diomedea exulans (Wandering Albatross).
        ,,   regia (Royal Albatross).
        ,,   chionoptera (Snowy Albatross).
    Thalassarche salvini (Grey-backed Mollymawk).
        ,,   melanophrys (The   ,,
        ,,   bulleri (White-capped   ,,
        ,,   culminatus (Grey-headed ,,
        ,,   chlororhynchus (Yellow-nosed Mollymawk).
    Phœbetria fuliginosa (Sooty Albatross).

13. O. Impennes: 282.
    Aptenodytes patagonica (King Penguin).
    Pygoscelis papua (Rock Hopper).
    Catarrhactes chrysocome (Tufted Penguin).
        ,,   pachyrhynchus (Crested Penguin).
        ,,   sclateri (Big Crested Penguin).
        · ,,   schlegeli (Royal Penguin).
    Megadyptes antipodum (Yellow-eyed Penguin).
    Eudyptula minor (Blue Penguin).
        ,,   albosignata (White-flippered Penguin).

14. O. Steganopodes: 298.
    Sula serrator (Gannet).
    Phalacrocorax carbo (Black Shag).
        ,,   varius (Pied   ,,
        ,,   sulcirostris (Little Black Shag).
        ,,   melanoleucus (Frilled   ,,
        ,,   brevirostris (White-throated Shag).
        ,,   punctatus (Spotted   ,,
        ,,   featherstoni (Chatham Island Shag).
        ,,   campbelli (Campbell   ,,   ,,
        ,,   colensoi (Auckland   ,,   ,,

Phalacrocorax chalconotus (Pink-footed Shag).
„ stewarti (Stewart Island „
„ onslowi (Pitt „ „
„ ranfurlyi (Bounty „ „
„ carunculatus (Rough-faced „
„ traversi (Macquarie Island „

15. O. Pygopodes: 319.

Podicipes cristatus (Crested Grebe).
„ rufipectus (Little „

16. O. Lamellirostres: 322.

F. Anatidæ: 322.
Casarca variegata (Paradise Duck).
Anas superciliosa (Grey „
Nettion castaneum (Grey Teal).
Elasmonetta chlorotis (Brown Duck).
Nesonetta aucklandica (Flightless Duck).
Spatula rhynchotis (Shoveller).

F. Fuligulidæ: 330.
Nyroca australis (White-winged Duck).
Fuligula novæ-zealandiæ (Black Teal).

F. Merganettidæ: 333.
Hymenolæmus malacorhynchus (Blue Duck).

F. Mergidæ: 335.
Merganser australis (Southern Merganser).

Sub-Class Ratitae: 337.

F. Apterygidae: 337.
Apteryx mantelli (Brown Kiwi).
„ australis (Southern Kiwi).
„ oweni (Grey Kiwi).
„ occidentalis (Spotted Kiwi).
„ haasti (Great Kiwi).

III. REPTILIA: 369.
    1. O. Lacertilia: 369.
        F. Geckonidæ: 369.
            Gehyra oceanica (Oceanic Gecko).
            Naultinus elegans (Spotted Lizard).
                  „   greyi (Green   „
                  „   rudis (Rough   „
            Dactylocnemis pacificus (Tree Lizard).
                  „   granulatus (Long-toed Lizard).
                  „   maculatus (Short-toed   „

        F. Scincidæ: 376.
            Lygosoma grande (Rock Lizard).
                  „   moco (Common   „
                  „   dendyi (Pitt's Island Lizard).
                  „   lineo-ocellatum (Long-tailed Lizard).
                  „   smithii (Short-tailed Lizard).
                  „   æneum (Copper   „
                  „   ornatum (Ornamented .,

    2. O. Rhynchocephalia: 380.
        Sphenodon punctatus (Tuatara).

IV. AMPHIBIANS: 384.
    O. Ecaudata: 384.
        F. Discoglossidæ: 384.
            Liopelma hochstetteri (N.Z. Frog).
                  „   hamiltoni (Stephen Island Frog).

NAME-LIST OF NEW ZEALAND BIRDS.

Mr. W R. B. Oliver, Dominion Museum, Wellington, has kindly supplied the following name-list of New Zealand birds, brought up to May, 1924.

In addition to names of birds that breed within the New Zealand geographical area, it includes the names of extra-limital species, that is, species which breed outside the New Zealand area, but occasionally visit New Zealand. On this account, the list is longer than the list in the main part of this work. In accordance with the present practice, the most highly specialised species, the Crows, are placed at the bottom of the list, and the lower forms at the top. The reverse order is retained in the main work, because it is not considered advisable in a semi-popular publication, now in its fourth edition, to alter that order.

The nomenclature of New Zealand birds still is in a confused state, and alterations in this list, probably, will be made. The names of the genera following, and their limits, so far as they are common to Australia and New Zealand, are those accepted in the "Official Check List of Australian Birds," which will be issued soon by the Royal Australasian Ornithologists' Union. Through the courtesy of Dr. J. A. Leach, they have been utilised in compiling the following list. The families are those of Messrs Mathews and Iredale's lists quoted below.

Family Spheniscidæ.

Thick-billed Penguins.

Eudyptula minor (Forster)	Little Blue Penguin
„ albosignata (Finsch)	Silver Penguin
Eudyptes pachyrhynchus (Gray)	Victoria Penguin
„ sclateri (Buller)	Big-crested Penguin
„ chrysocome (Forster)	Crested Penguin
„ schlegeli (Finsch)	Royal Penguin
Megadyptes antipodes (Hombron and Jacquinot)	Yellow-crested Penguin

Family Aptenodytidæ.

Thin-billed Penguins.

Pygoscelis papua (Forster)	Rock-hopper
Aptenodytes patagonica (Miller)	King Penguin

Family Thalassidromidæ.

Storm Petrels.

Oceanites oceanicus (Kuhl)	Yellow-webbed Storm Petrel
Garrodia nereis (Gould)	Grey-backed Storm Petrel
Pealea lineata (Peale)	Samoan Storm Petrel
Pelagodroma marina (Latham)	White-faced Storm Petrel
Fregata tropica (Gould)	Black-bellied Storm Petrel

Family Procellariidæ.

Shearwaters and Petrels.

Puffinus assimilis (Gould)	Allied Shearwater
„ gavia (Forster)	Forster's Shearwater
„ bulleri (Salvin)	Ashy-backed Shearwater
„ pacificus (Gmelin)	Wedge-shaped Shearwater
„ griseus (Gmelin)	Sooty Shearwater, or Mutton Bird
„ tenuirostris (Temminck and Laugier)	Short-tailed Shearwater
„ carneipes (Gould)	Fleshy-footed Shearwater
Procellaria parkinsoni (Gray)	Black Petrel
„ æquinoctialis (Linné)	White-chinned Petrel
„ cinerea (Gmelin)	Grey Petrel
Priocella antarctica (Stephens)	Silver-grey Petrel
Pterodroma macroptera (Smith)	Grey-faced Petrel
„ neglecta (Schlegel)	Kermadec Petrel
„ externa (Salvin)	Sunday Island Petrel
„ inexpectata (Forster)	Mottled Petrel
„ lessoni (Garnot)	White-headed Petrel
„ cookii (Gray)	White-winged Petrel
„ parvirostris (Peale)	Pacific Petrel
Daption capensis (Linné)	Spotted Petrel
Thalassoica antarctica (Gmelin)	Antarctic Petrel
Halobæna cærulea (Gmelin)	Blue Petrel
Pachyptila belcheri (Mathews)	Thin-billed Prion
„ desolatus (Gmelin)	Dove Prion
„ turtur (Kuhl)	Fairy Prion
„ vittata (Gmelin)	Broad-billed Prion
Macronectes giganteus (Gmelin)	Giant Petrel

Family Pelecanoididæ.

Diving Petrels.

Pelecanoides urinatrix (Gmelin)	Diving Petrel

Family Diomedeidæ.

Albatrosses.

Phœbetria fusca (Hilsenberg)	Sooty Albatross
„ palpebrata (Forster)	Light-mantled Sooty Albatross
Diomedea exulans (Linné)	Wandering Albatross
„ epomophora (Lesson)	Royal Albatross
„ melanophrys (Temminck and Laugier)	Black-browed Mollymawk
„ chinoptera (Salvin)	Snowy Albatross
„ bulleri (Rothschild)	Snares Island Mollymawk
„ chrysostoma (Forster)	Flat-billed Mollymawk
„ chlororhynchus (Gmelin)	Yellow-nosed Mollymawk
„ cauta (Gould)	Bounty Island Mollymawk
„ salvini (Rothschild)	Grey-backed Mollymawk

Family Fregatidæ.

Frigate Birds.

Fregata minor (Gmelin)	Frigate Bird
„ ariel (Gould)	Lesser Frigate Bird

Family Phalacrocoracidæ.

Shags, or Cormorants.

Phalacrocorax carbo (Linné)	Black Shag
„ ater (Lesson)	Little Black Shag
„ chalconotus (Gray)	Bronze Shag
„ campbelli (Filhol)	Campbell Island Shag
„ carunculatus (Gmelin)	Rough-faced Shag
„ varius (Gmelin)	Pied Shag
Stictocarbo punctatus (Sparrman)	Spotted Shag
„ featherstoni (Buller)	Chatham Island Shag
Microcarbo melanoleucus (Vieillot)	White-throated Shag

Family Anhingidæ.

Darters.

Anhinga novæhollandiæ (Gould)   Australian Darter

Family Pelecanidæ.

Pelicans.

Pelecanus conspicillatus (Temminck)   Australian Pelican

Family Sulidæ.

Gannets.

Sula leucogaster (Boddaert)	Brown Gannet
„ dactylatra (Lesson)	Masked Gannet
„ serrator (Gray)	The Gannet

Family Phæthontidæ.

Tropic Birds.

Phæthon rubricauda (Boddaert)   Red-tailed Tropic Bird

Family Sternidæ.

Terns and Noddies.

Chlidonias leucoptera (Temminck)	White-winged Tern
„ albistriata (Gray)	Black-fronted Tern
Hydroprogne caspia (Pallas)	Caspian Tern
Sterna striata (Gmelin)	White-fronted Tern
„ vittata (Gmelin)	Sub-Antarctic Tern
„ bergii (Lichstenstein)	Crested Tern
„ fuscata (Linné)	Sooty Tern
„ nereis (Gould)	Fairy Tern
Procelsterna cerulea (Bennett)	Little Grey Noddy
Anous minutus (Boie)	White-capped Noddy
Gygis alba (Sparrman)	White Tern

Family Laridæ.

Gulls.

Larus dominicanus (Lichstenstein)	Black-backed Gull
„ novæhollandiæ (Stephens)	Red-billed Gull
„ bulleri (Hutton)	Black-billed Gull

Family Stercorariidæ.

Skuas.

Catharacta lonnbergi (Mathews)	Great Skua
„ maccormicki (Saunders)	South Polar Skua
Stercorarius parasiticus (Linné)	Arctic Skua

Family Scolopacidæ.

Snipe, Sandpipers, etc.

Gallinago hardwicki (Gray)	Australian Snipe
„ aucklandica (Gray)	New Zealand Snipe
„ pusilla (Buller)	Chatham Islands Snipe
Calidris canutus (Linné)	Knot
Erolia testacea (Vroeg)	Curlew Sandpiper
Pisobia acuminata (Horsfield)	Sharp-tailed Stint
„ ruficollis (Pallas)	Red-necked Stint
„ minuta (Leisler)	Little Stint
„ maculata (Vieillot)	Pectoral Sandpiper
Glottis nebularius (Gunnerus)	Greenshank
Heteroscelus incanus (Gmelin)	Grey Sandpiper
Limosa lapponica (Linné)	Godwit
„ limosa (Linné)	American Godwit
Mesoscolopax minutus (Gould)	Little Whimbrel
Numenius phæopus (Linné)	Australian Whimbrel
„ cyanopus (Vieillot)	Australian Curlew
Crocethia alba (Pallas)	Sanderling

Family Phalaropidæ.

Phalaropes.

Phalaropus fulicarius (Linné)	Grey Phalarope

Family Recurvirostridæ.

Avocets and Stilts.

Himantopus leucocephalus (Gould)	Pied Stilt
„ novæzealandiæ (Gould)	Black Stilt
Recurvirostra novæhollandiæ (Vieillot)	Red-necked Avocet

Family Hæmatopodidæ.

Oyster-catchers.

Hæmatopus ostralegus (Linné)	Pied Oyster-catcher
„ unicolor (Forster)	Black Oyster-catcher

Family Arenariidæ.

Turnstones.

Arenaria interpres (Linné)	Turnstone

Family Charadriidæ.

Plovers.

Pluvialus dominicus (Muller)	Lesser Golden Plover
Charadrius obscurus (Gmelin)	New Zealand Dotterel
„ bicinctus (Jardine and Selby)	Double-banded Dotterel
„ ruficapillus (Temminck and Laugier)	Red-capped Dotterel
„ veredus (Gould)	Oriental Plover
Anarhynchus frontalis (Quoy and Gaimard)	Wry-billed Plover
Thinornis novæseelandiæ (Gmelin)	Shore Plover

Family Vanellidæ.

Wattled Plovers.

Lobibyx novæhollandiæ (Stephens)	Spur-winged Plover

Family Glaræolidæ.

Pratincoles and Coursers.

Stiltia isabella (Vieillot)	Australian Pratincole

Family Rallidæ.

Rails.

Rallus muelleri (Rothschild)	Auckland Island Rail
Cabalus modestus (Hutton)	Little Chatham Island Rail
Hypotænidia philippensis (Linné)	Buff-banded Rail
Nesolimnas dieffenbachii (Gray)	Chatham Island Rail
Galliralus australis (Sparrman)	Brown Weka or Woodhen
„ brachypterus (Lafresnaye)	Black Weka or Woodhen
„ hectori (Hutton)	South Island Weka or Woodhen
Crex crex (Linné)	Corncrake
Porzana pusilla (Pallas)	Marsh Rail
„ plumbea (Gray)	Swamp Rail

Family Gallinulidæ.

Gallinules.

Porphyrio melanotus (Temminck)	Pukeko or Swamp Hen
Notornis hochstetteri (Meyer)	Takahe
Tribonyx ventralis (Gould)	Black-tailed Water Hen

Family Fulicidæ.

Coots.

Fulica atra (Linné)	Coot

Family Podicipidæ.

Grebes.

Podiceps cristatus (Linné)	Great Crested Grebe
„ rufopectus (Gray)	Dabchick

Family Apterygidæ.

Kiwis.

Apteryx australis (Shaw and Nodder)	Kiwi
„ owenii (Gould)	Little Grey Kiwi
„ haastii (Potts)	Great Grey Kiwi

Family Perdicidæ.

Partridges and Quail.

Coturnix novæzealandiæ (Quoy and Gaimard)	New Zealand Quail

Family Treronidæ.

Fruit Pigeons.

Hemiphaga novæseelandiæ (Gmelin)	Wood Pigeon
„ chathamensis (Rothschild)	Chatham Islands Pigeon

Family Anatidæ.

Ducks and Geese.

Dendrocygna eytoni (Eyton)	Whistling Duck
Casarca variegata (Gmelin)	Paradise Duck
Anas superciliosa (Gmelin)	Grey Duck
Virago castanea (Eyton)	Grey Teal or Green-headed Teal
Elasmonetta chlorotis (Gray)	Brown Duck
Nesonetta aucklandica (Gray)	Auckland Islands Duck
Spatula rhynchotis (Latham)	Shoveller Duck
Hymenolaimus malacrorhynchus (Gmelin)	Mountain Duck
Nyroca australis (Eyton)	White-eyed Duck
Fuligula novæseelandiæ (Gmelin)	Scaup
Mergus australis (Hombron and Jacquinot)	Auckland Islands Merganser

Family Ardeidæ.

Herons and Bitterns.

Ardea cinerea (Linné)	Grey Heron
Notophoyx novæhollandiæ (Latham)	White-fronted Heron
Egretta alba (Linné)	Great White Heron
Demigretta sacra (Gmelin)	Blue Heron
Nycticorax caledonicus (Gmelin)	Night Heron
Ixobrychus minutus (Linné)	Little Bittern
Botaurus poiciloptilus (Wagler)	Black-backed Bittern

Family Plegadidæ.

Ibises.

Plegadis falcinellus (Linné)	Glossy Ibis

Family Plataleidæ.

Spoonbills.

Platalea regia (Gould)	Royal Spoonbill

Family Falconidæ.
Hawks.

Nesierax novæseelandiæ (Gmelin)   Quail-hawk
Cerchneis cenchroides (Vigors and
                  Horsfield)   Nankeen Kestrel

Family Aquilidæ.
Eagles and Harriers.
Circus approximans (Peale)   Harrier

Family Strigidæ.
Owls

Ninox novæseelandiæ (Gmelin)   Morepork
Sceloglaux albifacies (Gray)   Laughing Owl

Family Strigopidæ.
Owl-Parrots.
Strigops habroptilus (Gray)   Kakapo

Family Platycercidæ.
Broad-tailed Parrakeets.
Cyanoramphus novæzelandiæ (Spar-
                mann)   Red-fronted Parrakeet
      „   unicolor (Lear)   Antipodes Island Parrakeet
      „   auriceps (Kuhl)   Yellow-fronted Parrakeet
      „   malherbei (Souance)   Orange-fronted Parrakeet

Family Nestoridæ.
Nestor meridionalis (Gmelin)   Kaka
   „   notabilis (Gould)   Kea

Family Cuculidæ.
Cuckoos.
Cuculus optatus (Gould)   Oriental Cuckoo
Lamprococcyx lucidus (Gmelin)   Shining Cuckoo

Family Eudynamytidæ.
Koels.
Urodynamis taitensis (Sparmann)   Long-tailed Cuckoo

Family Coraciidæ.
Rollers.
Eurystomus orientalis (Gmelin)   Roller

Family Dacelonidæ.
Wood Kingfishers.
Halcyon sanctus (Vigors and
             Horsfield)   Kingfisher

Family Micropodidæ.

Swifts.

Micropus pacificus (Latham)   White-rumped Swift
Chætura caudacuta (Latham)   Spine-tailed Swift

Family Acanthisittidæ.

Riflemen.

Acanthisitta chloris (Sparmann)   Rifleman

Family Xenicidæ.

Island Wrens.

Traversia lyalli (Rothschild)   Stephen Island Wren
Xenicus longipes (Gmelin)   Bush Wren
  ,,   gilviventris (Pelzeln)   Rock Wren

Family Hirundinidæ.

Swallows.

Hylochelidon nigricans (Vieillot)   Australian Tree Swallow

Family Muscicapidæ.

Flycatchers and others.

Myiomoira macrocephala (Gmelin)   South Island Tomtit
  ,,   toitoi (Lesson)   North Island Tomtit
Miro australis (Sparmann)   South Island Robin
  ,,   longipes (Lesson)   North Island Robin
  ,,   traversi (Buller)   Black Robin
Rhipidura flabellifera (Gmelin)   Pied Fantail
  ,,   fuliginosa (Sparmann)   Black Fantail

Family Acanthizidæ.

Thornbill Warblers.

Gerygone igata (Quoy and Gaimard)   Grey Warbler
  ,,   albofrontatus (Gray)   Chatham Island Warbler

Family Campophagidæ.

Cuckoo Shrikes.

Graucalus robustus (Latham)   Little Cuckoo Shrike

Family Turnagridæ.

Island Thrushes.

Turnagra tanagra (Schlegel)   North Island Thrush
  ,,   capensis (Sparmann)   South Island Thrush

Family Bowdleridae.

Fern Birds.

Bowdleria punctata (Quoy and
                 Gaimard)   Fern Bird
  ,,   rufescens (Buller)   Chatham Island Fern Bird

Family Paridæ.

Tits and others.

Mohoua ochrocephala (Gmelin)   Yellowhead, or Bush Canary
„   albicilla (Lesson)   Whitehead, or Bush Canary
Finschia novæseelandiæ (Gmelin)   Creeper

Family Zosteropidæ.

White Eyes.

Zosterops lateralis (Latham)   White-eye, or Silver-eye

Family Meliphagidæ.

Honey-eaters.

Anthornis melanura (Sparmann)   Bellbird
„   melanocephala (Gray)   Chatham Island Bellbird
Notiomystis cincta (Du Bus)   Stitchbird
Prosthemadera novæseelandiæ
                    (Gmelin)   Tui
Acanthochæra carunculata (Latham)   Yellow Wattlebird

Family Motacillidæ.

Wagtails and Pipits.

Anthus novæseelandiæ (Gmelin)   Ground Lark

Family Heteralochidæ.

Huias.

Heteralocha acutirostris (Gould)   Huia

Family Creadiontidæ.

Saddlebacks.

Creadion carunculatus (Gmelin)   Saddleback

Family Callæadidæ.

Wattled Crows.

Callæas cinerea (Gmelin)   Orange-wattled Crow
„   wilsoni (Bonaparte)   Blue-wattled Crow

REFERENCES TO NOMENCLATURE

Following is an incomplete reference list of works dealing with the nomenclature of New Zealand Birds:—

Hutton, F. W.—"Index Faunæ Novæ Zealandiæ," 1904.

Buller, Sir Walter—"Supplement" to "A History of the Birds of New Zealand," 1905.

Mathews, G. M.—"Birds of Australia," in progress.

Mathews, G. M., and Iredale, T.—"A Reference List of the Birds of New Zealand," "Ibis," April, 1913, and July, 1913.

Mathews, G. M., and Iredale, T.—"A Manual of the Birds of Australia," in progress.

Benham, W. B.—"Nomenclature of the Birds of New Zealand," "Transactions N.Z. Institute," Vol. XLVI., 1914.

L. M. Loomis—"Review of the Albatrosses, Petrels, and Diving Petrels," "Proceedings Californian Academy of Science," 4th series, Vol. II., Pt. ii., No. 12, San Francisco, April 22, 1918.

Ridgway, R.—"Bulletin U.S. National Museum," No. 50, Pt. viii., Washington, 1919.

Bent, Arthur Cleveland—"Life Histories of North American Petrels and Pelicans and their Allies," "Bulletin U.S. National Museum," No. 121, Washington, 1922. (Deals with habits.)

Mathews, G. M., and Iredale, T.—"A Name-list of the Birds of New Zealand," "Austral Avian Record," Vol. IV., 1920.

Printed by Whitcombe and Tombs Limited—G37138